THE NATURE OF MATTER

THE NATURE OF MATTER

PHYSICAL THEORY FROM THALES TO FERMI

GINESTRA AMALDI

TRANSLATED BY
PETER ASTBURY

The University of Chicago Press
Chicago and London

Originally published in Italy as
Materia e Antimateria
© 1961 by Arnoldo Mondadori Editore

The University of Chicago Press, Chicago 60637
©1966 by George Allen & Unwin Ltd.
All rights reserved. Published 1966
Phoenix edition 1982
Printed in the United States of America

89 88 87 86 85 84 83 82   5432
LCN: 66-12133

ISBN: 0-226-01661-7

TO THE MEMORY OF MY PARENTS

INTRODUCTION

When, twenty-five centuries ago, the Greek philosopher Thales of Miletus asked himself, 'What is the world made of and how is it made?' he put a question which has not yet been answered. We are not even certain that there is a definite answer.

We know today that all matter existing in the universe is made up of atoms; but the study of the central part of this atom—the part which is called the atomic nucleus—is still far from complete. Some people even think that the study of the constitution of matter can never reach a final conclusion; for this would not only lie outside the scope of our instruments; it would also lie beyond the range of our imagination. A scientist has written: 'The universe is not only stranger than we suppose; it is perhaps stranger than we can suppose.'

But this does not interest physicists, who continue to plan and carry out experiments and to construct theories. They well know that the object of research is not to give final answers, nor to discover the true essence of things; it is only to study phenomena and to find the laws which they obey: that is to construct a system into which known phenomena can be fitted, and into which new phenomena will logically fall as they come to be discovered.

The work of physicists, until recently regarded as too scientific and arid, is now followed with very great interest. But this work has always been, and is now more than ever before, closely linked with complex mathematical theories which invoke profound scientific understanding.

I would like in this book to expound the work of atomic physicists, in a form as simple as possible. I will show how, starting from the first speculations of the Greek philosophers, man has arrived at the most up-to-date theories of the atomic nucleus, at the great conquest of nuclear energy and at the discovery of new particles, whose role in the architecture of the world is not yet in any way explained. I will tell of the long path which has up till now been followed, of the existing problems which the physicist of today is trying to answer and of the prospectives for the science of tomorrow.

CONTENTS

INTRODUCTION   *page* 7

I. ATOMS AND MOLECULES   17
   The Greek Philosophers—The Atomic Theory of Leucippus and of Democritus—The Value of the Theory of the Greek Philosophers—The Atomic Theory after the Renaissance—Lavoisier and Proust—Dalton and the Atomic Theory—Molecules and Atoms; Avogadro—Cannizaro—Atomic Weight—The Periodic Law of the Elements—The Masses and Dimensions of Molecules and of Atoms—The Kinetic Theory of Matter

II. THE ATOM   45
   The Electron—The Atomic Model of J. J. Thompson—The Discovery of Radioactivity—Alpha Rays, Beta Rays and Gamma Rays—The Beta and Gamma Rays—Alpha Rays—Rutherford's Collision Experiments—The Atomic Nucleus—The Atomic Model of Rutherford—Atomic Physics and Nuclear Physics

III. ATOMIC PHYSICS   68
   Ionization—New Difficulties—Quanta—Bohr's Hypotheses—The Correspondence Principle—Sommerfeld's Conditions—Probability—The Hypothesis of the Spinning Electron—Pauli's Exclusion Principle—Value of the Bohr-Sommerfeld Theory—New Quantum Theories—Light—The Photo-Electric Effect—The Wave Theory and the Corpuscular Theory of Light—Wave Mechanics—De Broglie Waves—Heisenberg Mechanics—The Uncertainty Principle—Causality and Uncertainty—The Complementarity Principle—Overcoming the Wave-Particle Paradox—Conclusion

IV. NATURAL RADIOACTIVITY   114
   Isotopes—Radioactive Transformations—The Radioactive Families—The Laws of Radioactive Disintegration—Radioactive Equilibrium—Radioactive Substances in Nature—The Age of the Rocks—The Internal Heat of the Earth—Cosmic Rays

V. HOW RADIOACTIVE PARTICLES ARE DETECTED 130
The Passage of Charged Particles through Matter—The Passage of Photons through Matter—Methods of Detection of the Particles—Counters—Track Detectors

VI. THE ATOMIC NUCLEUS 157
Nuclear Physics—The Masses of Atoms—The Discovery of Nuclear Transformations—Discovery of Neutrons—Protons and Neutrons in the Nucleus—Neutron Proton Diagram—The Mass Defect of a Nucleus—The Binding Energy—The Positron—Creation and Annihilation of Electron-Positron Pairs—Nuclear Forces—Nuclear Models—Emission of Gamma Rays—Alpha Disintegration—Beta Disintegration—The Neutrino—Fermi's Theory—The Hypothesis of the Existence of the Meson

VII. ARTIFICIAL RADIOACTIVITY 195
Artificial Transmutations—Nuclear Reactions—Various types of transmutation.—Artificial Radioactivity—Bombardment by Neutrons—Slow Neutrons—Radioactivity Produced in Heavy Elements

VIII. ACCELERATING MACHINES 212
Acceleration of Particles—Accelerating Machines

IX. COSMIC RAYS 226
The Penetrating Radiation—The Intensity of Cosmic Rays as a Function of Altitude—The Latitude Effect—The Constitution of the Primary Radiation—Where do Cosmic Rays Originate?—The Theory of Fermi—The Journey of Cosmic Rays Through Space—Cosmic Radiation in the Atmosphere—The Meson in the Cosmic Radiation—The μ Meson and the π Meson—Cosmic Ray Phenomena in the Atmosphere

X. NUCLEAR ENERGY 248
The Binding Energy of Nucleons—Exoenergetic and Endoenergetic Reactions—Fusion and Fission—Fusion or Thermo-Nuclear Reactions—Plasma—Energy Losses in the Plasma—The Magnetic Bottle—The Pinch Effect—The Energy of the Stars—Fission—Chain Reactions—Uranium—Nuclear Reactors—Types of Reactor—Atomic Fission Bombs—Effects of Atomic Explosion—The Peaceful Uses of Nuclear Energy

XI. SUB-NUCLEAR PARTICLES 283
New Developments in Physics—What Do We Mean by 'Elementary Particles'?—The First Particles—The Spin of Elementary Particles—Statistics—Inter-Actions Between Particles—The Anti-Particles—The Pion and the Neutrino—Twelve Particles—The Muon—Strong Interactions and Weak Interactions—New Strange Particles—Associated Production of Strange Particles—The Law of Conservation of 'Strangeness'—The Principle of Parity—Weak Interactions and Neutrino Physics—Future Questions—The Anti-Particles and the Stability of Matter—Still More Particles and Anti-Particles—Anti-Matter—Does Anti-Matter Exist in the Universe?—Conclusion

PLATES

1. *An X-ray photograph* facing page 64
2a. *Tracks of alpha particles in a Wilson chamber with magnetic field* 65
  b. *Tracks of beta rays in a Wilson chamber with magnetic field* 65
3a. *Tracks of protons and electrons in a Wilson chamber with magnetic field* 96
  b. *Tracks of protons in a bubble chamber* 96
4a. *Microphotograph of the disintegration of an atomic nucleus produced in a special photographic emulsion by a high energy particle (track 1 on the right hand side of the photograph)* 97
  b. *Pairs of electrons and positrons produced by gamma rays of 335 MeV* 97
5. *Spark chamber in a magnetic field* 224
6a. *The inside of a linear accelerator* 225
  b. *The 1·5 metre constant frequency cycletron at Argonne (USA)* 225
7a. *The synchrotron at Frascati near Rome* 256
  b. *Internal view of the ring of the 25 Gev proton synchrotron at Cern, Geneva* 256
8. *External view of the Ispra reactor* 257

FIGURES

1. The Greek theory of the four elements *page* 19
2. Two volumes of hydrogen combine with one volume of oxygen to make two volumes of water 31
3. Cathode ray tube 46
4. Canal ray tube 50
5. Path of alpha, beta and gamma rays in an electric field 56
6. Wavelength 57
7. Wave-lengths of electromagnetic radiations 58
8. Rutherford and Royds' apparatus 59
9. Rutherford's experiment 62
10. Deflection of an alpha particle in the case of Thomson's atomic model and in the case of Rutherford's model 63
11. Radioactive transformation from radium into radium-C' 119
12. The three radioactive families 120
13. Compton effect 133
14. Ionization chamber 136
15. How an ionization chamber is connected 136
16. Crystal ionization chamber 138
17. A Geiger counter 138
18. Arrangement of three counters in coincidence 140
19. Apparatus for the detection of a particle which originates in a lead screen 141
10. A spintariscope 142
21. The Cerenkov effect 144
22. The Wilson chamber 147
23. A Wilson chamber in coincidence 149
24. The first experiment to show the possibility of making a bubble chamber; the bulb on the left, containing liquid ether, was warmed in a bath to 285°C; the thin tube between the two bulbs contained ether vapour. When the bath on the left was withdrawn the pressure

Figures

in the tube fell from 21 to 1 atmosphere and the liquid ether in the right-hand bulb became superheated: it boiled as soon as a radioactive substance was brought near it. 154

25. Diagram showing the principle of a spark chamber. Alternate plates are connected electrically to A. When a particle passes through counters C_1 and C_2 a high-voltage pulse is applied to A. The ionization in the track of the particle leads to sparks being formed along the trajectory. 155
26a. Hydrogen 169
26b. Deuterium 169
26c. Tritium 169
27. Proton-neutron diagram 171
28. Sketch of the photograph which enabled Anderson to discover the positron 174
29. The nuclear force 187
30. Whilst a neutron is not affected by the electric field of a nucleus, a charged particle, to enter a nucleus, must overcome the potential barrier due to its electric charge 196
31. Sketch of a constant voltage linear accelerator 216
32. Sketch of the betatron 219
33a. Sketch of the cyclotron 221
33b. Path of the particles in a cyclotron 222
34. Intensity of cosmic rays as a function of height 228
35. Trajectories of a positive particle and of a negative particle deflected by the earth's magnetic field 230
36. Binding energy per nucleon as a function of atomic weight 250
37. The pinch effect 262
38. Types of magnetic bottle used in the study of fusion reactions 264
39. Sketch of a nuclear reactor 275
40. Orientation of the axis of a participle in an external magnetic field 289
41. Revolving particle and mirror image 305
42. The experiment of Wu and her collaborators 306

CHAPTER ONE

ATOMS AND MOLECULES

THE GREEK PHILOSOPHERS

The first of the Greek natural philosophers was Thales of Miletus who lived in Ionia (Asia Minor) between 624 and 548 BC; he was, as Plato said, 'an able craftsman' and the originator of 'many skilful discoveries in the arts and in other activities'.

Thales was an engineer, statesman, navigator and surveyor and was astute in business affairs. According to tradition it was with him that Greek philosophy started; for he tackled the problem of the nature of things and was the first to endeavour to give an explanation based on the observations of the senses.

Thales asked himself what was primeval matter, the only matter which lasts for ever and which renews itself, which is the beginning of everything and whose transformations give rise to all the phenomena which man observes in the universe; what is, in other words, the matter from which all real things are formed? He gave the answer that the beginning of everything is water because, as Aristotle was to say later, seeds germinate in dampness and everything on which they nourish themselves is damp: it is water which gives rise to ice and which in the alluvial deposits of rivers consolidates itself into land; it is water which evaporating becomes air; and finally it is water which comes forth from the earth in the streams and which falls in rain from the skies. And so water is the primeval element from which all other things originate.

This intrepid idea of the existence of a primeval element, this idea that qualitatively matter is unitary constitutes the principal merit of Thales who in this way started twenty-five centuries ago that investigation of nature which later became one of the main fields of study of the Greek philosophers.

Anaximander and Anaximenes carried on the Ionic school of

Miletus. For Thales the primeval substance was water. On the other hand Anaximander (fellow citizen and friend of Thales who lived between 610 and 547 BC) claimed in a curious theory that everything was formed from the 'apeiron', the infinite, the undetermined. In the description of Theofrastes: 'Anaximander declares that it is not air nor water nor any other of the so-called elements but a substance different from these which is infinite and from which are generated the skies and the world between them.'

As Aristotle explained, the infinity of matter is due to its being the matrix of eternal becoming. 'One sees then,' writes Enriques, 'that in the spirit of Anaximander there appears the tendency to move out from our world, to free oneself from every boundary, finding everywhere a type of diffuse matter . . . which could not only be quantitatively infinite but also infinitely diffused.'

In the succeeding trials this attribute of primeval matter will be maintained. There will be proposed other primeval substances but they will no longer be too concrete and limited; there will be a matter which satisfies the conditions of infinity; fire for Heraclitus and air for Anaximines.

But, if there exists a substance which is unitary, eternal, impenetrable and uniformly spread which makes up everything which exists and which as the Pythagorians affirmed is void of quality and has only the properties of existing and filling uniformly all space, then everything that our senses tell us is illusion. The flow of things, their continuous transformation, is an illusion. The rationalist Parmenides of Elea (born about 520 BC), whom Plato called 'venerable and formidable', could no longer find in this universe the explanation of a change or of any future whatsoever. According to a pseudo-Aristotelian work 'without boundary in space and in time and everywhere similar the One (the World) is immobile'. And Parmenides himself affirmed 'and so birth vanishes and death is inconceivable'. In this way thought, repudiating the evidence of the senses, constructs a world which is as a whole without movement, without future and without life.

The logic of the Greek rationalist philosophers arrived then at the negation of reality; and it arrived at this by starting from the premise of a unitary substance (the Being) which had no quality other than existence and extension and from the premise of the inconceivability of the vacuum; for, as Melixes was later to say, 'vacuum is nothing and nothing (the Not Being) cannot exist'. Starting from these premises it arrived by logical methods at the absurd. It was necessary therefore to modify the premises.

There were then three ways open to philosophers. The first, although maintaining the premise that the vacuum does not exist, takes as its starting point the hypothesis that primeval matter is not unitary, that it has different components and that these are such that the matter which we know results from their mixing together in various proportions. And so Empedocles of Agrigento (who lived from 490 to 430 BC) claimed that four substances formed 'the root of everything'; fire, air, earth and water. And Anaxagoras (who lived between 500 and 428 BC) said that matter was the result of a mixture of infinite qualities which cannot be seen but which exist unchanged: damp and dryness, cold and heat, dark and light etc.

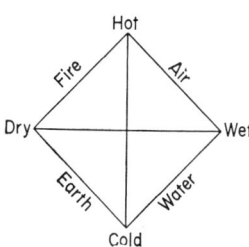

FIG. 1. The Greek theory of the four elements.

Another way of overcoming the difficulty raised by the rationalist philosophers of Elea started from the premises that the primeval substance is certainly unitary, devoid of quality and impenetrable (as Parmenides had claimed); but it is not compact: it is sub-divided into many parts of various sizes and of different forms which move through the void. As Aristotle writes in Book One of his *De Generatione et Corruptione*, for these philosophers '... this *plenum* is not one but many things of infinite number, and invisible owing to the minuteness of their bulk. These are carried along in the void ... and when they come together, they cause coming-to-be and when they dissolve, they cause passing-away.' Each one of these masses is complete, that is indivisible: and from the Greek word 'atomos' which means 'indivisible' they take the name of 'atoms'. We shall return soon to this atomic theory of matter of the early Greek philosophers.

Let us first conclude this very rapid and schematic review of the different philosophical schools which struggled against the rationalism of Elea—the rationalism which had led to those paradoxes to which I have just referred. The third way does not start as the two previous ways from a modification of some of the premises which, when rationally developed, had led to the conclusion of Parmenides of Elea and of other philosophers of this school; it attacks and refutes the Elean theory from a more general point of view—on the philosophical ground of the theory of knowledge.

The Elean philosophers had arrived at the negation of development and of movement; they had distinguished between the 'opinion' of men and the rational 'truth'; they had, that is, claimed that man falls into error when he relies on sense perception: our senses tell us that things are multiple and changeable but we discover through reason that the Being is single and unchangeable. A stand against this negation of the reality which is displayed to us by our senses was made by the first Greek sophists Protagoras and Gongias (both born towards 485 BC). They claimed that all knowledge is acquired by man by means of his senses; one should not therefore search for the nature of things in themselves but they should be regarded in their relations with the human world: existence is not, as Parmenides had affirmed, a property of the object in itself; it is relative to the man who by means of his senses is in contact with it. As Plato makes Socrates say in the *Theaetetus*: '... that which is perceived must become so to someone, when it becomes sweet or bitter or the like; for to become sweet, but sweet to no one, is impossible."

In this way the sophists criticized the rationalist concept of science; it was a criticism which later attacked also the social order, the customs, the rights and the religious principles of Greek society in the fifth century before Christ.

These then were the various reactions of philosophers to the rationalism of the Elean school.

THE ATOMIC THEORY OF LEUCIPPUS AND OF DEMOCRITUS

The atomic theory which was, as I have said, put forward as an attempt to overcome the difficulties into which the Elean philosophy had been led was proposed by Leucippus.

Almost nothing is known about the life and the works of this

Atoms and Molecules

philosopher. He was certainly a contemporary of Empedocles (490–430 BC) and was perhaps born at Mileto; but there is no document about his doctrine or about his personality because his writings were incorporated in the works of his great disciple, Democritus, who was an older contemporary of Socrates (470–399 BC) and who lived for a very long time: ninety years or, according to others, very emphatically 109 years.

It is known that the atomic theory was proposed by Leucippus and was then taken up and developed by Democritus. But with very few exceptions historians do not distinguish between the ideas of the two philosophers; following them I myself in these short notes will present as a single whole the doctrine of the master and of his disciple; I will restrict myself to expounding their atomic concept of matter and their kinetic theory of the world and neglect other problems (of knowledge and of ethics) which Democritus treated fully in the construction of his system.

Democritus was born in the Greek colony of Thrace. He was very wise and enormously versatile; he understood the scientific thought and philosophy of his predecessors and of his contemporaries. Aristotle says of Democritus: 'it seems that he has thought about everything'. A catalogue of his written works, which has been passed on by Diogenes, attributes to him works on physics, logic, the theory of knowledge, mathematics, astronomy, music, literary questions, medicine, botany, zoology, oriental questions and on the technique of war. Even if some of these works have been attributed wrongly to Democritus, as some scholars maintain, it is certain that this philosopher, who lived as a sage and who is said to have spent the whole of his patrimony on voyages of instruction, passionately dedicated his long life to the study of nature. It was he who said, 'I would give the crown of the king of Persia for a scientific discovery'; it is he who, according to Levi, 'surpasses Aristotle in the width of his scientific interest, in his love of empirical research and in the range of his literary production'.

According to the atomic theory of Leucippus and of Democritus, who developed the hypothesis of a single primeval element, everything is made up of real individual entities (the atoms) and of a vacuum. Admitting the existence of the vacuum the atomists boldly arrayed themselves against one of the fundamental theses of Parmenides. He had said, 'the vacuum, the non-Being cannot exist'; Democritus claimed, 'the Nothing is something just like the Being'.

Atoms are all made of the same substance and differ amongst themselves in their size and in their shape. They are indivisible not because of their extreme smallness (though they must be very small because they cannot be perceived in matter), but because they do not contain vacuum; as Cicero was later to say, an atom is indivisible *propter solidatem*. Like the Being of Parmenides atoms are immutable, are not created and are eternal and possess no sensible quality. Democritus himself wrote: 'colour is a convention, sweetness is a convention, bitterness is a convention; in reality there exist only the atoms and the void'; that is, the sensible attributes of things (colour, sweetness and bitterness . . .) exist only subjectively because they are relative to the senses of the man who perceives them.

And so the origins of everything are the fullness (the atoms) and the vacuum. The atoms are made of identical material but differ in their sizes and shapes; they continually move in all directions through the vacuum; this movement is not created and is eternal; and when the atoms group themselves together in a whirlpool they form things and when they separate from each other the things are dissolved. Thus things change (and only in appearance are they created and destroyed) thanks to the continuous movement of the infinite number of atoms in the infinite vacuum. In this way are formed not only water, air, fire and earth but also all complex substances; and infinite worlds are infinite because of the infinite number of atoms and because space is infinite.

This is the theory held by the Greek atomists; for them, the scope of science is not, as it was for the rationalists, pure and rational knowledge (truth) as opposed to the knowledge which is obtained from the senses (opinion) which is only an illusion; it is not as it was for the sophists only the knowledge obtained from the senses; but as noted by Plato in the *Theaetetus*, for the atomist science is 'true opinion accompanied by reason'. Change is not denied but comes to be considered as the result of the physical relations between immutable entities.

The doctrine of Leucippus and Democritus was then an attempt to interpret the world presented by the senses by means of mechanistic concepts. But although it formed the basis for other philosophical systems which followed, it nevertheless did not flourish. Until the Renaissance the qualitative physics of Aristotle was to be dominant.

THE VALUE OF THE THEORY OF THE GREEK PHILOSOPHERS

When the Greek philosophers confronted the great problem of the nature of things they put forward theories which today appear arbitrary because they are not based on any experimental foundation. But these theories are extremely valuable: for the first time man tried to find an explanation of the world free of preconceived ideas and superstitious beliefs. The Ionic philosophers were the first to recognize that the world is not, as Schrödinger has written, 'a stage on which spirits and gods act according to the impulses of the moment and in a more or less arbitrary fashion . . . but it is something which could be understood if someone dedicates himself to observing it attentively. . . .' These philosophers had the immense merit of being curious and knowing how to wonder and we know that these are two fundamental characteristics of the scientific attitude.

Furthermore all the theories on the constitution of matter upheld by the Greek philosophers of the fifth century before Christ display one common concept: the idea that in all the transformations of matter and behind the infinite variety of the things about us, there exists something which remains immutable.

In this way the Greek spirit, with a power of imagination which it is today difficult to appreciate in its full value, arrived at the first concept of that which later developed into what we call today the 'scientific method': to try to explain the complex and apparently exceedingly diverse phenomena which nature offers us by means of a few simple phenomena which obey precise, general laws.

But if the rôle of science is to seek the laws of nature the weapon of the scientists since Galileo has been the experimental method: the scientist must conceive and execute experiments which allow him to test the validity of the laws which other experiments have allowed him to enunciate. But when one asks how it is possible that the atomic theory of matter, which had been proposed and developed by Leucippus and by Democritus in the fifth century before Christ, should have had to wait two millenia before being re-established we should not forget that the Greeks lacked experimental science as we understand it today: their theories were abstract.

The Greeks who discovered the beauty of pure thought did not recognise the value of experimental test. This is perhaps due, amongst other things, to social causes. In Greek society manual labour was unworthy of free men and was entirely carried out by the slaves; a scholar thought, made theories and discussed. But it was unthinkable that he should, with materials and tools, dedicate himself to a manual task; this was work for a slave.

THE ATOMIC THEORY AFTER THE RENAISSANCE

The hypothesis that matter was made up of extremely small and indivisible particles was then enunciated for the first time more than four hundred years before the birth of Christ. Attacked and discredited by Aristotle, modified by Epicurus (342–270 BC), and in the poetry of *De Rerum Natura* expounded in this modified form by Lucretius (95–55 BC), it was brought to life again after the Renaissance.

After it had been upheld by Giordano Bruno (1584–1600) it acquired in the first half of the seventeenth century that mechanistic character which it had previously lost. This was due above all to the work of Francis Bacon (1561–1625) who, although criticizing the doctrine of Democritus, supported a corpuscular theory; and of Galileo (1564–1642) who accepted the atomic theory and outlined it briefly but clearly.

Father Gassendi (1592–1655) revived the old theory of Epicurus and, it seems, was the first to recognize that many of the properties of gases can easily be explained by the movement of the very small particles of which they are constituted. Isaac Newton (1642–1727) taking up the ideas of Gassendi wrote in his *Optiks*: '... it seems probable to me, that God in the Beginning form'd Matter in solid, massy, hard, impenetrable, moveable Particles ... and that these primitive Particles being Solids, are incomparably harder than any porous Bodies

Atoms and Molecules

compounded of them; even so very hard, as never to wear or break in pieces ...', and Voltaire in his *Dictionaire Philosophique*, writes: 'the solid is today considered as a chimera ... the void is admitted; the hardest bodies are considered to be as full of holes as a sieve and are in fact so. Atoms are accepted —indivisible and immutable'.

In this way the atomistic hypothesis was maintained in life through the centuries. It was only in the earliest years of the nineteenth century that it imposed itself as necessary for the interpretation of experimental facts. On the one hand an atomic theory was formulated in order to account for the laws which govern physical phenomena; and on the other hand, based on the hypothesis of the atomic constitution of matter, mechanical interpretation of thermal phenomena led to the formulation of the kinetic theory of gases.

It is necessary to underline a conceptual difference between the atomic doctrine of the ancient Greeks and modern scientific atomic theory: this difference is due to a completely different mental outlook on the problems posed by nature. The object of the Greeks, as of other ancient peoples, was to synthesize the whole of knowledge in a single formula; it was to find a doctrine which would explain all phenomena; a true scholar should be able to reply to all questions. On the other hand, the atomic theory put forward at the beginning of the nineteenth century certainly did not have the ambitious object of supplying a universal interpretation of everything. It is accepted only as a hypothesis which is able to account for a certain restricted number of facts discovered by experiment. In particular, in order to explain other properties of matter other facts continued to be interpreted by non-corpuscular hypotheses: electricity, heat, the force of gravity were all entities which maintained a continuous character. When, later, new electrical phenomena were to be discovered they could not be interpreted according to the hypothesis that electricity is a continuous fluid; there was then formulated a corpuscular theory of electricity in which the old and the new phenomena could logically be inserted.

A Greek scholar should have been able to reply to any demand. The modern scientist, who proceeds by checking hypotheses by experiments, dedicates himself on the other hand to the study of well defined and circumscribed problems. He should be endowed with particular qualities; he should be, amongst other things, modest and sincere: modest because he knows that he cannot arrive at the truth only by thought, but that this must be checked, and kept on the right road by means of experiment; sincere because he must honestly accept the results of experiment even if these results contradict everything which he had previously accepted.

LAVOISIER AND PROUST

By the end of the eighteenth century chemists began to have a precise idea of what constitutes a pure, chemically identifiable substance; they began to distinguish simple bodies, that is those which cannot be decomposed by chemical means (the *elements*), from compound substances; and they already proposed that the number of simple bodies is very small compared with the enormous multitudes of compound bodies.

The eighteenth century had been a period of intense development in pure mathematics. There had been put forward mathematical theories (in which a corpuscular structure had been assigned to matter) of certain properties of matter, such as adhesion, cohesion and capillary phenomena. The theory of the movement of the planets, which had been elaborated by Newton, served as a model for the new theories; in Newton's theory the corpuscles of matter are considered as massive points which obey the laws of mechanics: the new theories were therefore mechanical and corpuscular. Thus Laplace (1749–1827) elaborated a theory of capillary phenomena; and in 1738 Bernouli (1700–81) proposed the kinetic theory of gases to which I shall return later; with this theory there were rediscovered in an exact form the gas laws which had been revealed by experiment.

These atomistic and mechanical ideas had in the eighteenth century profoundly modified and re-created in a new form ideas about the constitution of matter and its transformation. In chemistry Lavoisier (1743–94) had furnished science with a fundamental instrument for further progress: the *law of the conservation of matter* during a chemical reaction. He had enunciated this law in plain words in his elementary *Treatise on Chemistry* which was published in 1787: 'Nothing is created in the operation either of art or of nature and one can pose as a principle that in every operation there is an equal amount of matter before and after the operation; that the quality and the quantity of the initial substance is the same and there are only changes and modifications.'

In this way there was introduced into chemistry the concept of quantitative measurement; and chemists equipped with balances applied themselves to the study of the weights of chemical compounds. In 1799, in the course of a famous, placid and courteous polemic with Berthollet which lasted eight years Proust (1755–1826) enunciated the *law of constant proportions*, in which he affirmed that a chemical compound always contains the same elements combined together in the same proportions: for example, in the formation of water (however it may be obtained) any given weight of hydrogen combines always with eight times the same weight of oxygen.

Proust said to his opponents: 'we cannot create compounds as we please. When you believe that you can combine bodies' (i.e. the elements) 'in arbitrary proportions you, short-sighted wretches, are only making mixtures of which you are incapable of distinguishing the parts; what you are making are monsters...' and elsewhere he affirmed: 'a compound is a substance to which nature assigns fixed ratios; it is, in short, a being which nature never creates otherwise than with balance in hand.'

DALTON AND THE ATOMIC THEORY

This time knowledge was sufficiently advanced to allow the transformation into a scientific theory of what had previously been only an hypothesis—the hypothesis that matter is made up of atoms. This important step was made by John Dalton, an English chemist who was born in 1766 and who died in 1844.

Patient research into the composition of gases led him to enunciate another quantitative law of chemistry—*the law of multiple proportions*—and the atomic theory. Historians of science are not agreed on the order of these two discoveries: some say that he first conceived the atomic theory and that in order to verify it he then carried out those researches which led to the discovery of the law of multiple proportions; others on the other hand claim that it was the discovery of this law which led him to elaborate the atomic theory appropriate to the new experiments. In any case on October 21, 1803, before nine members of the Literary and Philosophical Society of Manchester of which he was Secretary, Dalton read a paper in which for the first time he expounded the fundamental principles of his atomic theory. He then explained it in his lectures and five years later, in his book *A New System of Chemical Philosophy*, showed how the atomic hypothesis allowed one to account for the law of Lavoisier and the law of Proust and how it constituted an incomparable instrument for the systemization of chemical phenomena.

Dalton clearly saw in the atomic theory the essential principle of all the chemical laws concerning weight. The characteristic of his atomic theory which differentiates it from previous atomic hypotheses is the new criterion which formed the base on which it was developed: for the first time there entered into chemistry the idea of the *weight* of atoms.

According to the atomic theory of Dalton matter is made up of homogeneous and exceedingly small particles: the *atoms* which conserve their individuality in all chemical processes. Every chemical element (that is, every simple substance which cannot be further decomposed by chemical means—such as hydrogen, oxygen, sulphur, iron, gold, etc.) is made up of a particular type of atom; the atoms in a given element are all equal to each other: in particular they have all the same invariant weight. Different elements have atoms of different weights. Each element therefore is characterized by the weight of its atoms.

Dalton himself wrote in his *New System of Chemical Philosophy*: 'Chemical analysis and synthesis go no further than to the separation of particles from one another, and to their reunion. No new creation or destruction of matter is within the reach of chemical agency. . . . All the changes we can produce, consist in separating particles that are in a state of cohesion or combination, and joining those that were previously at a distance.

'In all chemical investigations, it has justly been considered an important object to ascertain the relative *weights* of the simples which constitute a compound. But unfortunately the enquiry has terminated here; whereas, from the relative weights in the mass, the relative weights of the ultimate particles, or atoms of the bodies might have been inferred.... Now it is one great object of this work, to show the importance and advantage of ascertaining *the relative weights of the ultimate particles, both of simple and compound bodies, the number of simple elementary particles which constitute one compound particle, and the number of less compound particles which enter into the formation of one more compound particle.*'

And further on he added:

'if two bodies A and B are disposed to combine together these are the combinations which can take place starting from the simplest:

1 atom of A + 1 atom of B = 1 atom of C, binary
1 atom of A + 2 atoms of B = 1 atom of D, tertiary
2 atoms of A + 1 atom of B = 1 atom of E, tertiary
1 atom of A + 3 atoms of B = 1 atom of F, quaternary'

In this way, with marvellous sobriety, Dalton showed how the atomic hypothesis justified the Lavoisier law of the conservation of mass, Proust's law of constant proportion and also his own law of multiple proportions; this states that: 'When two elements combine to form more than one compound the different weights of one which combine with the same weight of the other stand to each other in ratios of simple and integral numbers'; that is they are related to each other as one:two: three.... When for example oxygen combines with nitrogen it can give rise to five different oxides of nitrogen: in these oxides the weight of oxygen which combines with a hundred parts of nitrogen are respectively

$$57; 114; 171; 228; 285$$

these weights are related to the other as 1:2:3:4:5 (as one can see by dividing all of them by the first, that is by 57).

This was the enormous contribution made to science by the orderly and logical mind of Dalton. There are those who do not agree in assigning to Dalton the dominant position in the story of scientific thought attributed to him by the majority of historians; they assign it rather to Lavoisier and they affirm that after his discovery of the principle of the conservation of matter it was inevitable that the formulation of the atomic theory of matter would be arrived at; the bases for this theory had moreover been introduced after Lavoisier by other scientists who came before Dalton. But this is inevitable in the march of time: at a certain moment the totality of the experimental evidence is such that the discovery of a new theory which accounts for it in a logical manner is, we may say, in the air. And this is demonstrated by the fact that not infrequently in the history of science the same discovery is often made independently, but at the same time, by more than one research worker. It is however fair to attribute great merit to he who at a certain moment, with logical intuition, is able to clarify and co-ordinate into a new theory the results and the tentative suggestions of those who have gone before him.

MOLECULES AND ATOMS; AVOGADRO

Dalton then established the bases of the atomic theory; but it was necessary to finish his work. This was to wait for Amedeo Avogadro.

Dalton had accepted that an element is made up of extremely minute particles of matter (*atoms*), all equal to each other and that the different elements are made up of atoms with different weights; but he also accepted that elements in the free state (that is, when they are not combined) are all made up of single atoms: that hydrogen, for example, was made up of single atoms of hydrogen, oxygen of single atoms of oxygen, etc.; and that compounds on the other hand are made up of *molecules*, each of which is made up of the atoms of the component elements: thus, a molecule of water would be made up of an atom of hydrogen and an atom of oxygen. For Dalton the concept of a molecule referred only to compounds whilst for the elements the expression 'molecule' and 'atom' were equivalent.

But with this hypothesis there soon arose an insurmountable difficulty: a new law was discovered in 1808 by Gay-Lussac (1778–1850) who sustained that in chemical reactions involving gaseous substances the volumes of the elements which interact and the volumes of the products which are obtained (all

Atoms and Molecules

measured at the same pressure and at the same temperature) are related to each other by ratios which can be expressed by simple integral numbers. Well, according to the hypothesis of Dalton, this law is shown to be in irresolvable disagreement with the already known result of fundamental experiment.

In 1811 the Italian Count Amedeo Avogadro (1776–1856) perfected and completed Dalton's theory by supplying to it, as Avogadro himself said, 'a new means of defining the link which we have found with the general fact established by Gay-Lussac'.

Considering the results of Gay-Lussac Avogadro reached the conclusion that if one thinks of a gas as a collection of elementary particles of *molecules* independent one from the other one must accept that 'in identical conditions of temperature and of pressure equal volumes of gas, simple or compound, contain the same number of molecules'; by 'molecule' he meant the smallest mass of a substance which is capable of independent existence. This statement has been given the name of *Avogadro's Law*.

In this way Avogadro clearly distinguished the concept of molecule from the concept of atom; and he, as opposed to Dalton, considered that gaseous elements are *not* all made up of single atoms; but that each of them (hydrogen, oxygen, chlorine, nitrogen ...) are like compounds made up of molecules: in the elements the molecules are formed of identical atoms, whilst the molecules of compounds are formed of different atoms and in particular of the atoms of the elements which enter into the combination.

This, for example, is how we can represent the combination of hydrogen (whose molecule is made of two atoms: H_2) and of oxygen (whose molecule is also made up of two atoms: O_2) to make water (H_2O):

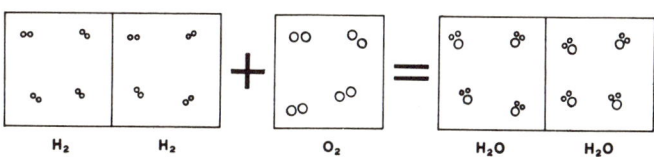

FIG. 2. Two volumes of hydrogen combine with one volume of oxygen to make two volumes of water.

So the molecules of the elements can be formed of a single atom, or of two or of three, etc. identical atoms; for example, the molecule of potassium is *monatomic*, that is, formed of a single atom; those of hydrogen, of oxygen, of chlorine, etc. are *diatomic*, that is, formed of two atoms, etc.

Thus even for the simple elements one can distinguish between atoms and molecules.

The addition of this important complement to Dalton's ideas effectively enabled chemistry to build a coherent system of formulae for the constitution of bodies.

CANNIZARO

As Avogadro himself said, his law supplied 'a method of determining the relative masses of the atoms and the proportions in which they enter into combinations.'

But the enunciation of Avogadro's Law was greeted with indifference and incomprehension; he returned again to his theory in later publications; but scientists (apart from a few very rare exceptions) paid him no attention.

But when they set to work to determine the *relative* atomic weight of the elements that were then known (taking by convention as the unit the atomic weight of hydrogen, the lightest of the gases) they encountered serious difficulties which arose just from the fact that they hardly ever took account of Avogadro's Law. For more than fifty years these difficulties accumulated until an intolerable situation was reached: there was a chaotic confusion of formulae; as Mayer has noted the same organic formula might indicate three different substances, according to whether the reader followed one theory rather than another; and in organic chemistry there were three different tables of atomic weight with corresponding differences in the formulae of compounds, in the mechanism attributed to reactions, etc. The development of different branches of chemistry became continually more and more entangled in the difficulties which arose from this lack of a criterion in the choice of atomic weights.

Scientists began to look sceptically on the atomic hypothesis and Dumas, head of the French school of chemistry, reached the point of declaring that if it were in his power he would have cancelled the word 'atom' from the scientific vocabulary.

This confusion continued until 1858 when the young Italian chemist, Stanislao Cannizaro, fortunately took up again the atomic theory and showed how, starting from Avogadro's definition of the molecule it was possible to arrive in a simple and direct manner at the determination of the atomic weights; he took as the atomic weight of an element the smallest amount of the element which can enter into the constitution of a molecule. Cannizaro expounded his ideas first in various publications and later in 1860 at the International Chemistry Congress which was held at Carlsruhe in Germany, for the express purpose of attempting to reach agreement on the uncertain and involved question of atomic weight.

Cannizaro had made himself champion of Avogadro's ideas and succeeded in bringing them to the attention of the Congress. And though agreement was not immediately reached, and a further forty years had to pass before the polemics provoked by the atomic idea were finally concluded, nevertheless from that time onwards most of the chemists recognized and accepted the clarificatory work of the young Cannizaro.

The concept of molecule was thus clearly differentiated from the concept of atom. And the success of the atomic hypothesis in the interpretation of the vast amount of material assimilated by chemists in a century of work provided for this hypothesis a very solid foundation.

ATOMIC WEIGHT

Dalton had directed his attention to the determination of the *relative atomic* weights, taking as a unit the weight of the lightest atom existing in nature: the atom of hydrogen. The atomic weight of an element is then the number which gives the ratio of the weight (or of the mass) of an atom of that element to the weight (or of the mass) of an atom of hydrogen. Dalton published a table which supplied the relative atomic weights of twenty-one substances.

Struck by the fact that the atomic weight of oxygen is a whole number and by other similar regularities the Englishman Prout put forward in 1815 a hypothesis which has been defined as 'a revolutionary generalization'; according to this hypothesis the atomic weight of all the elements are integral multiples of that of hydrogen. In 1816 Prout claimed that the simplest explanation of this observed regularity is to suppose that the atoms of all elements are formed by the combination of a smaller or larger number of atoms of hydrogen. Hydrogen would then be the 'primeval matter' or 'prote yle' (from the Greek 'prote' = first and 'hule' = matter).

Chemists became very fascinated by this simple hypothesis. But as the determination of the atomic weight became more and more accurate the hypothesis became more and more criticized. Stas claimed that 'it is only an illusion, a pure hypothesis definitely contradicted by experiment'. Only later was interest reawakened in Prout's hypothesis; it will reappear but in a new form.

As I have said above the determinations of the first atomic weights were made by taking as equal to one the atomic weight of hydrogen; and it had been found that with this unit the atomic weight of oxygen was equal to 16. But more accurate measurements very soon disclosed that this value was slightly too high: by about 1%. Later the unit for measurement of atomic weight was changed. Instead of taking as equal to one the atomic weight of hydrogen the atomic weight of oxygen was fixed as equal to 16; thus the unit was taken as a sixteenth part of the atomic weight of oxygen. This choice was made, although it may seem strange and inconvenient, for two sound reasons: first of all because the determination of the atomic weight of hydrogen is particularly delicate and subject to error since hydrogen is the lightest of the elements; and further it is particularly suitable that the atomic weight of oxygen should be expressed by a whole number given that it enters into combinations suitable for exact analysis with most of the other elements.

This convention was maintained until recently.[1] When the atomic weight of oxygen is taken equal to 16 the atomic weight of hydrogen becomes slightly greater than one: it is equal to 1·008; but on the other hand the atomic weight of a large number of elements then become so close to a whole number that in many calculations chemists can use the whole number without making serious errors.

[1] In 1961 carbon replaced oxygen as the standard.

THE PERIODIC LAW OF THE ELEMENTS

When chemists had determined the atomic weights of a sufficiently large number of elements they were offered (as a shrewd chemist has said) a just reward: the discovery of the periodic law of elements.

In 1864 the Englishman Newlands '... announced the existence of a simple relation or law among the elements when arranged in the natural order of their atomic weights, to the effect that the eighth element, starting from a given one, was a sort of repetition of the first, or that the elements belonging to the same group stood to each other in a relation similar to that between the extremes of one or more octaves in music.' And he proposed to designate this simple relation by the provisional term 'law of octaves'. For example, if one writes the elements in order of ascending atomic weight lithium, sodium and potassium occupy respectively the third, the eleventh and the nineteenth place; now lithium, sodium and potassium are from a chemical point of view similar elements.

But this relation was not entirely satisfactory; it was very coldly received and the Chemistry Society of London refused to publish Newlands' work in its *Journal*.

In Newlands' table there is contained the germ of one of the most important laws of chemistry (more important than was apparent when it was first put forward); but there is no doubt that there should be awarded to the Russian chemist Dimitri Mendeleev the merit of having clearly established the relation between the properties of the elements and their atomic weight, of having brought this result to the attention of the chemists of his time and of having placed it as the basis of a comprehensive system of classification.

In this way Mendeleev himself wrote in his book *Principles of Chemistry*: 'there must be some bond of union between mass and the chemical elements; and as the mass of a substance is ultimately expressed ... in the atom, a fundamental dependence should exist ... between the individual properties of the elements and their atomic weight. But nothing ... can be discovered without looking and trying. So I began to look about and write down the elements with their atomic weights ... and this soon convinced me that *the properties of the elements are in periodic dependence upon their atomic weight*; and although I have had my doubts about some obscure points, yet I have never once doubted the universality of this law, because it could not possibly be the result of chance.'

Mendeleev published his law in 1869. He was the first to try to bring together all the elements in a single system which is called the *Periodic System of Elements*; it is given in a modern form in Table I.

In the Periodic System the ninety-two elements which exist in nature are arranged in eight vertical columns, called *groups*, each group being sub-divided into two columns, *a*, *b*. These groups are obtained by sub-dividing suitably into *periods* the continuous series of elements arranged in ascending order of their atomic weights. The ordinal number of the elements in the Periodic System is called the *atomic number*; thus, the element of atomic number 1 is hydrogen, that of atomic number 92 is uranium, whose atom is the heaviest of the atoms existing in nature.

The Periodic System of Elements is then an arrangement in which the chemical elements are disposed in ascending order of atomic weight and in which at regular intervals there re-appear elements which have similar chemical behaviour (for example, lithium, sodium, potassium or chlorine, bromine and iodine, etc.) in such a way that when arrayed in a suitable table they appear in the same column. In this way there stand out clearly the analogies and the differences between all the elements.

The Periodic Table was immediately confirmed by very many results. First of all Mendeleev had modified the atomic weights of certain elements which were already known in 1869 in order to put them in the place in the table which belonged to them according to their chemical properties; these modifications were confirmed by subsequent measurements. There then remain in the table certain empty places; from their position in the table Mendeleev could predict the chemical properties of the elements which should have occupied these empty spaces; in particular he predicted the existence of three new elements (which he called eka-boron, eka-aluminium and eka-silicon), which were immediately searched for, found and baptized with the names of scandium, gallium, and germanium. Finally, after their

TABLE I.—Mendeleev's Table. For every element there is shown, in order: the atomic number, the atomic weight of the mixture of isotopes which occurs in nature and the numbers of the various natural isotopes. Where these do not appear it means that the element does not exist in nature. For the last two rows see also Table IV.

discovery, the noble gases were found places in the Periodic System.

The Periodic System triumphantly closed and summarized the period of research which was started with Lavoisier. We shall later see its profound significance—a significance which neither Mendeleev nor his contemporaries could expect.

THE MASSES AND DIMENSIONS OF MOLECULES AND OF ATOMS

So chemistry allows one to determine precisely the relative atomic weights of the elements; that is, to express the weight of the atoms of different elements in definite units derived from one of them (the atom of oxygen). It allows one then to determine the *ratios* between the weights of various atoms; but it cannot predict the mass in grammes of one atom. Furthermore chemistry claims that the molecule and the atom are so small that they escape from direct observation as individuals without giving any indication of their size.

But very soon certain phenomena made possible the determination of the mass in grammes of single molecules and thence of single atoms.

Already in 1827 a young Scottish doctor, Robert Brown (1733–1858), had made a discovery which remained however for a half century without explanation. Brown examined in a microscope a drop of liquid in which there were suspended exceedingly minute grains of pollen; he observed that these moved about continuously and randomly, each one following a zig-zag path independently of the path followed by the neighbouring particle; this is a movement which never stops. It has taken from its discoverer the name of *Brownian motion*.

This discovery did not attract attention. Nevertheless Brownian motion is extremely strange. It is observed in all fluids which carry in suspension very small particles and is the more active the less viscous is the fluid and the smaller the particles in suspension. It can be observed for years without signs of diminishing; it is observed in liquid which has been enclosed for thousands of years in quartz; it is spontaneous and eternal.

These characteristics of the Brownian motion led C. Wiener to conclude in 1863 that 'this agitation originates neither in the

Atoms and Molecules

particles nor in some cause external to the liquid but must be attributed to internal movements which are characteristic of the fluid state'. The conclusions of Wiener were later clearly completed and formulated; in 1905 Einstein provided for the Brownian motion a quantitative theory which was in 1906 completed by Smoluchowski.

The disorderly movement of particles in suspension is due to the collisions which these undergo with the molecules of the liquid, molecules whose movement is the more active the higher is the temperature of the liquid. Brownian motion is then intermediate between ourselves and the world of the molecule: the particles in suspension are large enough to be observed in the microscope but they are also small enough to be kept in movement by the collisions with the molecules of the fluid in which they are immersed.

This claim which was then only a hypothesis was subjected in 1908 to experimental proof by the Frenchman, Jean Perrin (1870–1942), in a series of famous experiments for which he was awarded in 1926 the Nobel Prize for Physics.

This first proof of the real existence of molecules made it possible to establish the value in grammes of the masses of molecules, values which were later confirmed by measurements carried out with other independent methods. The finding in this way of results which agreed within the limits of experimental errors provides the best proof of the reality of the atomic hypothesis.

Before recording the masses of molecules and of atoms and their dimensions a small parenthesis is necessary in order to introduce a practical method of writing in figures these enormous numbers (and also these extremely small numbers). One can well understand that to write for example, the number 90 000 000 000 000 is not only inconvenient but also introduces the danger (it is especially great when a number enters into an arithmetical operation) of writing one or two extra zeros or too few.

An extremely simple method is therefore used. It is only necessary to remember that a hundred equals ten squared, a thousand equals ten to the three, ten thousand equals ten to the four ... that is that multiples of ten are equal to a power which has ten as its base and the number of zeros as its exponent.

Then the number which I have written just above is written very much more simply as 9×10^{13}.

For very small numbers, that is for numbers with many zeros after the decimal point, a similar method is used. Since $0·1 = 1/10 = 10^{-1}$: $0·01 = 1/100 = 10^{-2}$: $0·001 = 1/1000 = 10^{-3}$, etc., one can say that every sub-multiple of ten can be written in the form of a power which has ten as base and as *negative* exponent the number of zeros which precede the significant figure *including* the zero which precedes the decimal point. For example, the number 'one hundredth, thousandth, millionth' can be written more simply as 10^{-11} instead of 0·00000000001; and the number 'seven hundredth thousandth millionths' will be 7×10^{-11}.

Let us close now the parenthesis and return to the mass of the atom.

It has been found that the lightest atom which exists, the atom of hydrogen, has a mass of $1·673 \times 10^{-24}$ gm., i.e. 1·673 billionths of a billionth of a gramme; this means that a collection of about 603 million billion atoms of hydrogen would have the mass of a milligramme. The heaviest atom which exists in nature, the atom of uranium, has a mass of 398×10^{-24} grammes.

The size of a molecule or, what is equivalent, the number of molecules contained in a given volume of gas has been determined by making use of very different phenomena; it has been determined from the viscosity of gases, from the diffusion of light from the sun (which produces the blue colour of the sky), from radioactive phenomena; and these different methods have all led to the same result: in one litre of any gaseous substance there are contained 27,000 million billion molecules (that is 27×10^{21} molecules). Molecules are therefore exceedingly small: the dimensions of an inorganic molecule are of the order of a few times 10^{-8} centimetres (that is a few hundredth millionths of a centimetre).

Thus in the earliest years of our century there was obtained direct experimental verification of the atomic theory; molecules and atoms had been weighed and measured.

THE KINETIC THEORY OF MATTER

Physicists had reached the stage when they were able to

determine not only the dimensions of a molecule but also its movement and its velocity; by means of the *kinetic theory* of matter they were able to present a complete and detailed picture of the internal state of a gas; this theory arose from an experiment by Dalton on the diffusion of gases, and was enriched and perfected by Maxwell, by Boltzmann and later by Einstein.

According to this kinetic theory (from the Greek word kinema = movement) all the molecules which make up matter are not at rest but in continual movement, with a velocity which, as we shall see, is the greater the higher the temperature.

To understand this theory would imply a deep knowledge of the laws of gases and of the main concepts of thermodynamics. It is not therefore suitable for extensive explanation here; but it will be appropriate to trace at least the essential characteristics.

We have not yet asked ourselves why the atoms and molecules which make up substances remain combined together and various bodies do not spontaneously disintegrate as, for example, a sand castle would disintegrate. The answer lies in the fact that the atoms and molecules exert forces of interaction on the surrounding atoms and molecules. These forces, which are often given the name of *cohesion forces*, are, at least up to a certain point, the greater the closer the molecules are together.

We can then only say that if the forces of cohesion are strong compared to the energy with which according to the kinetic theory the molecules move, these forces will prevent the molecules from moving away from their initial position; each molecule will only be able to vibrate about an equilibrium position; this case corresponds to the solid state of matter. If on the other hand the forces of cohesion are small compared to the molecular energies then the various molecules can disperse and move freely about the body to which they belong; this case corresponds to the liquid and gaseous states of matter.

According to these ideas the change of state of a substance is interpreted as a change in the relation between the energy of molecular movement and the forces of cohesion. As an example of a substance which can exist in three states—solid, liquid and gaseous—let us consider the case of water. Suppose that a piece of ice is heated; it will melt and form water; if the heating is continued the water will start to evaporate and the vapour will leave the container in which the ice was placed. This means that when the temperature is below zero degrees the energy of cohesion overcomes the energy of molecular movement and the molecules will then remain bound about their equilibrium position (the solid state); when the temperature is increased the molecules acquire continuously greater energy so that they are able to overcome the forces of cohesion; they move with increasing velocity in the liquid which they cannot however abandon (the liquid state); when the temperature is still further increased, the molecules, as a result of the great energy which they acquire, finally free themselves from all the cohesive bonds and can leave at a high velocity in any direction, completely independent one from the other (the gaseous state).

We conclude that the energy and the associated velocity of the molecules is the greater the higher the temperature.

Let us pause to consider the case of a gas and look rather more closely at the movement of its molecules which travel in all directions with high velocities.

Despite its high velocity a molecule never succeeds in normal conditions in travelling uninterrupted for a long distance; it is continuously deviated by collisions against the other molecules of the gas; the result is that the molecules follow a zig-zag path, continuously colliding and bouncing away in a new direction.

A molecule collides not only with other molecules but also against the walls of the container. And because the number of molecules is enormous these collisions against the walls are so frequent that they can be considered as continuous. They reveal themselves as a continuous thrust which the gas exerts on the walls of the container: this is the *pressure* of the gas on the walls. The tyre of a bicycle or a motor car remains blown up, despite the weight which it carries, just because of these very frequent collisions which the molecules of the compressed air in the tyre exert on its walls.

Naturally the number of collisions of the molecules with each other is the greater the larger the number of the molecules and the higher their velocity. Thus the pressure of a gas increases with the increase of the number of molecules and with the increase of the temperature of the gas.

For a given concentration and for a given temperature the number of collisions between the molecules, which determines the *internal friction* of the gas, depends on their size; clearly the collisions are more frequent when the molecules are larger. Thus from measurement of the internal friction of a gas the size of the constituent molecules can be estimated.

This is how kinetic theory can be used to measure the size of a molecule.

One might think that for a quantitative investigation of the properties of a gas it would be necessary to follow the history of each individual molecule. This would be impossible; in fact every molecule undergoes more or less violent collisions which deflect it more or less abruptly but always in a random and disordered manner; the paths of the individual molecule are therefore extremely intricate. Despite this the laws which gases obey are simple and were determined with relative ease.

This was possible thanks to the enormous number of molecules which make up a gas. We know from statistics that when an average is made from a large number of individuals the irregularities are diminished and the greater the number of individuals the greater is the smoothing of the irregularities.

This is a general feature of large numbers, so that not only the statistics of a collection of disorderly molecules lead to simple and well-established laws but the same thing also occurs, for example, in the statistics of a population. If we consider a particular bodily characteristic, for example height, in successive generations, we will find a very large number of irregularities of an absolutely capricious nature. But if on the other hand we consider the variations in the average height of all the people in one country at different times the individual fluctuations disappear and are replaced by a law of regular and continuous change, which is determined by controllable elements, such as the changes in environment, in economic conditions, in hygiene, etc.

It was thanks to Maxwell and to Boltzmann that the kinetic theory of matter was enriched by statistical concepts.

Naturally it is unnecessary to detail the procedures followed to determine the characteristics of a molecule. I will say only that these characteristics were not obtained using only one method but by making use of various phenomena; we therefore have solid ground for believing that they are very close to reality.

As we already know the order of magnitude of the atomic radius is 10^{-8} centimetres. At normal temperatures a molecule of oxygen has a velocity of about 400 metres a second; that is, if it were not deflected by the other molecules that it encounters in its path, it would travel from London to Edinburgh in about half an hour. But this premise never obtains because, as we shall see, each molecule continuously encounters other molecules and collides with them, thus continually changing its direction.

The average distance travelled by a molecule without undergoing a collision is given the name of *mean free path*; it depends of course on the distance between the molecules and thus, in a gas, on its density. For a gas at atmospheric pressure the mean free path of the molecules is about 10^{-5} centimetres; this means that one of these molecules collides a hundred thousand times in a centimetre of path making a very complicated zig-zag journey; at the end of its journey it will be displaced from its original position by only three-hundredths of a millimetre. The molecule undergoes about four thousand million collisions in a second.

CHAPTER TWO

THE ATOM

THE ELECTRON

But in the meantime a strange thing had happened. Even before the real existence of atoms had been experimentally demonstrated it was known that if they did exist they could not be compact and indivisible; they had in their turn to be made up of smaller particles. The word atom had then already lost its original significance as an indivisible particle.

There were many physical phenomena which showed that there must be electrically charged particles inside the atom. In order of time the first of these phenomena to be observed was the electrolysis of solutions; then came the phenomena of the discharge of electricity through gases and the simultaneous development of the electromagnetic theory of light on the one hand and of spectroscopy on the other. All of this led, in little more than fifty years of intense and brilliant research, to the inescapable conclusion that there exist inside atoms a number of identical negatively charged particles. These particles are *electrons*.

In 1833 the Englishman, M. Faraday (1791–1867) had put forward his laws on electrolysis which show that—as Helmholz (1821–94) was later to point out at a conference held in London in 1881—'if one accepts the hypothesis that elementary substances are made up of atoms, one cannot escape the conclusion that electricity itself is divided into elementary parts which behave as atoms of electricity'. In 1874 Stoney gave the name of electrons to these elements of electric charge.

But this idea remained almost unchanged and in practice did not lead to any further development until 1879 when Crookes, who was conducting experiments on the passage of electricity through gases, took up the study of a phenomenon which

Plücker had observed in 1869, that is the phenomenon of *cathode rays*.

Let us take a glass tube (Fig. 3), which contains a very, very rarefied gas and into which are fused the two poles of a generator of electric current; the positive pole is called the anode and the negative pole the cathode. If we make current flow there will be an electrical discharge in the tube. One observes on the wall of the tube opposite the cathode and exactly aligned with it, a small fluorescent patch which disappears when the discharge stops. Clearly during the discharge there is a beam of rays which

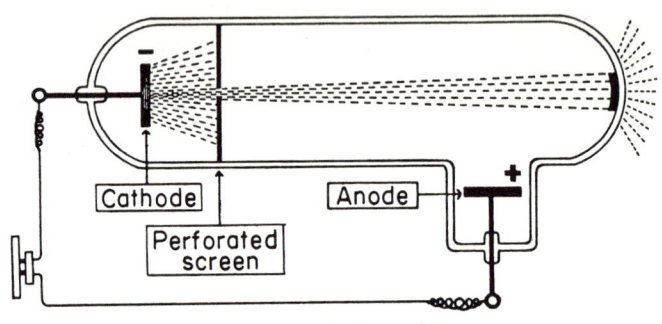

Fig. 3. Cathode ray tube.

leaves the cathode to strike the opposite wall and makes it fluoresce. As these rays come from the cathode they were given the name of *cathode rays*.

But what are these rays? Are they electromagnetic or are they particles? Are they, that is, rays of the same type as electromagnetic radiation (light, ultra violet rays, X-rays ...) or are they, on the other hand, a beam of particles which come from the cathode and collide at high velocity against the wall of the tube. After observations and experiments by various physicists the Englishman, J. J. Thomson (who was awarded the Nobel Prize in 1906) showed that cathode rays are made up of particles and particles which have a negative electric charge. These particles are those elementary charges whose existence had been proposed by Helmholtz and to which Stoney had given the name of electrons.

In 1897 there no longer existed any doubt about the existence of these negatively charged particles; 1897 can be considered as the year in which the electron was born.

The Atom

What is the mass of an electron? What is the value of its negative electric charge? In order to answer these questions very delicate and elegant experiments were made; these experiments have remained famous in the history of physics.

In a tube, such as that shown in Fig. 3, the beam of cathode rays travels in a straight line but if we put this tube in a magnetic field the electrons (because of their electric charge) are influenced by the magnetic field and will be deflected; the beam of cathode rays will now follow a curved trajectory. From observations on the deflection of the beam, under the influence of known magnetic fields one can arrive at the value of the ratio e/m between the charge of the electrons and its mass.

Now the charge of an electron was measured by Millikan; between 1908 and 1913 he developed to a high level of precision some older techniques and was thus able to carry out one of the most elegant experiments of modern physics.

The idea of the experiment is surprisingly simple. Let us imagine that we produce in air a spray of oil; the drops of oil in the course of being formed become charged with electricity. If watching, through a microscope, we fix our attention on a single drop, we see that it falls vertically under the action of gravity. We can, however, arrange an electric field so that the drop (being charged) will rise with a velocity which will naturally depend on the intensity of the applied electric field. By comparing the two movements of rising and falling one can calculate the ratio between the forces which are acting in the two cases and thence the value of the electric charge carried by the drop.

One finds that a drop can never have a charge smaller than a certain value; but its charge may be two, three or as much as eight times greater than this value. This elementary electric charge is just the value of the charge of an electron.

As one can see, this is an exceedingly delicate experiment. Millikan observed, for as long as an hour, the rising and falling of a single drop along a distance little greater than a millimetre.

If one knows in this way the electric charge of an electron one can immediately obtain the value of its mass from the known ratio e/m measured by means of the deflection of cathode rays under the influence of a magnetic field.

It was found in this way that the mass of an electron is about 1,800 times smaller than the mass of a hydrogen atom, which is the lightest known atom (i.e. an electron has a mass of 0.98×10^{-27} gm.). Its negative electric charge, again very small, has the value of 4.80×10^{-10} electrostatic units. A simple numerical example will give an idea of the smallness of this charge. The number of electrons which flow every second through the filament of an ordinary 100 watt electric light bulb is equal to the number of cubic centimetres of water which have flowed down the Tiber from the year 1200 until today.

Electrons can be obtained not only from a discharge tube but also by heating a metal or by irradiating it with ultra violet light; electrons are spontaneously emitted (as we shall see) from radio-active substances; there are electrons in the cosmic radiation; and finally huge showers of electrons of solar origin occasionally arrive in the earth's atmosphere.

'Once the nature of the electron was defined,' writes P. Chauson, 'technologists could make use of it and guide it in every possible form of electromagnetic field, directing it on to targets of X-ray tubes, making it draw, with Zworykin, a thousand arabesques on oscilloscopes and television tubes, applying to it the law of optics in modern electromicroscopes. Thus we are today at home with the electron.'

And now let us sum up. All electrons are identical. An electron has the smallest electric charge which exists in nature and is very much lighter than the lightest atom; it can be produced from any material, that is from any atom. It is reasonable to conclude that the electron is an essential constituent of all atoms.

The year in which the electron was born belongs to a series of four successive years which have been called 'the four golden years' which started the heroic age of physics. These are the years 1895–98. In 1895 X-rays were discovered. In 1896 the phenomenon of radioactivity, in 1897 the electron and in 1898 radium. These discoveries not only make a decisive turn in the march of science; they also provide physicists with new and powerful means of research.

THE ATOMIC MODEL OF J. J. THOMSON

The study of various phenomena has then shown that all atoms contain electrons. Electrons have, however, a negative electric charge whilst the atom in normal conditions is electric-

The Atom

ally neutral; it therefore follows that inside an atom there must be a positive electric charge and that the value of the positive charge must exactly balance the negative charge carried by the electrons.

One reaches in this way the conclusion that atoms, although they are indivisible by chemical means (that is, although they behave in all chemical senses as minute, indivisible and solid spheres) nevertheless have structures of their own in which positive and negative electric charge comes into play. Furthermore, the fact that the mass of electrons is very small compared with the mass of atoms suggests the hypothesis that the atomic mass is essentially connected with the positive charge.

The nature of the particles which carry the positive charge inside the atom was revealed by an experiment which Goldstein carried out in 1886, using a slight modification of the Crookes tube which we have discussed above; holes were made at various points in the cathode of the tube (Fig. 4).

When a discharge takes place in the tube one sees a narrow luminous beam emerge from each hole in the cathode. Thus while the cathode rays *leave* the cathode and go to collide against the opposite wall these new rays are formed in the gas which lies between the cathode and the anode and they travel to *collide against* the cathode; and if there are holes in the cathode they pass through them.

Because these rays are canalized in the holes of the cathode they have been called *canal rays*.

Canal rays, just like cathode rays, are made up of particles; but when they are in a magnetic field they are deflected in the opposite direction to the cathode rays. Thus the particles which make up the canal rays carry a positive electric charge.

But although the particles which make up the cathode rays, i.e. electrons, are always the same, that is, have the same mass and the same charge whatever the cathode may be and whatever gas may be contained in the tube, the positive particles which make up the canal rays have a mass and a charge which depends on the gas contained in the tube. The mass of each of these particles is rather less than, but almost equal to, the mass of the atom of the particular gas being studied and the charge of one of these positive particles has always a value which is a multiple of the electric charge carried by an electron.

These particles have then a positive charge and are very slightly lighter than atoms. These are the positively charged parts of the atom which must exist in order to balance the negative charge carried by the electrons.

Thus an atom of any element is made up of two parts. The first is a positively charged particle which has a mass which is almost equal to the mass of the atom and (which is therefore different from element to element). The second part consists of a number of identical electrons each of which has a mass which is almost negligible and a negative electric charge which is the smallest charge found in nature. The positive charge of the

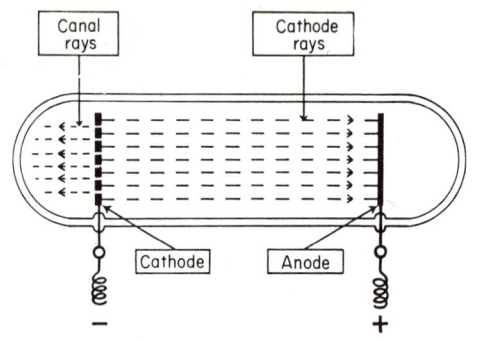

FIG. 4. Canal ray tube

first part is equal to the sum of the negative charges carried by the electron.

The problem then was to discover how the positive and negative charges were distributed in the atom. That is, to suggest a model of the atom which would allow one to account for all these observed phenomena and in particular which would allow one to account for the extraordinary stability of atoms.

In 1904 the Englishman J. J. Thomson put forward a hypothesis which turned out, however, to be inadequate to explain experimental facts. According to Thomson's hypothesis, an atom would be made up of a sphere of uniformly distributed positive electricity in which would be embedded the electrons like the seeds inside an apple; these electrons would be in equilibrium under the influence on the one hand of their neutral repulsions and on the other hand of the electrostatic attraction towards the centre of the positive sphere.

As I have said this model of Thomson's was abandoned because it was found to be in conflict with many experimental facts. The problem was resolved in 1911 by the New Zealand physicist Ernest Rutherford who, as a result of a famous experiment, put forward the model of the atom which even today forms the best picture that we can make of an atom. For this decisive experiment Rutherford used a technique which had been made available to physicists by the discovery of radioactivity which took place at the end of the last century.

We will, therefore, leave for a short time the problem of the constitution of the atom in order to discuss briefly the phenomenon of radioactivity.

THE DISCOVERY OF RADIOACTIVITY

It is often said that radioactivity was discovered by a lucky accident; but this is not strictly true. The discovery was made when its time was ripe and it was the result of happy intuition rather than precise method of work by the man who discovered it: the Frenchman Henri Becquerel.

The discovery of radioactivity happened in 1896; it lay therefore between the discovery of X-rays and the discovery of the electron; it threw open the door to a new world.

In January 1896 two months after the discovery of X-rays the French physicist Henri Becquerel (1862–1908) had the idea of investigating whether phosphorescent substances which had been excited by light emitted a radiation similar to X-rays. The salts of uranium were among the many other phosphorescent substances which he chose for his experiments; the particular choice of uranium salts was due to the fact that they had been the object of a study by himself and his father in previous years.

Becquerel took sheets of disulphate of uranium and of potassium and put them on a packet of thick black cards which contained a photographic plate, in such a way that the plate could not be affected by the phosphorescent light. This arrangement was left in the sun and then the plate was developed. This experiment, which was repeated several times, showed that after an exposure to sunlight of several hours, those parts of the plate which were underneath the sheet were very slightly darkened.

According to Becquerel's son there was no sun on February 26th and on the 27th it appeared only intermittently. Whilst waiting for a better light the packet was shut up in a drawer. The sun re-appeared on March 1st. Henri Becquerel was about to expose the packet to the sun when he had second thoughts and took it to the dark-room. It would be better to change the plates because, as the uranium salts had been exposed to the diffuse light of the 26th and for a short period of sunlight on the 27th before being put into the dark, the experimental conditions were not precisely defined. Furthermore the possibility that a slight impression had been obtained should not be neglected. One of the plates was immediately developed. The result was extraordinary: the impression was very much stronger than in previous experiments. It was clear that irradiation had been emitted even in the absence of light. It was the discovery of radioactivity. In this historic photograph one sees the radiograph of a thin copper cross which had been placed under one of the two sheets.

Uranium salts then, even if they are not excited by light, spontaneously emit a penetrating and invisible radiation which is able to make an impression on a photographic plate, even after passing through thin metal sheets. That is, making use of a word which was subsequently introduced, uranium is 'radioactive'.

Immediately after the discovery of this phenomenon a young woman took up the investigation as her subject for a thesis. She was an intelligent Polish student, Marie Sklodowska, who had come to Paris to study science and who had in 1895 married Pierre Curie, a lecturer at the Sorbonne. The life of this scientific couple, a simple life of untiring work, appears surrounded by a romantic halo.

They were united in a perfect scientific collaboration and united in the desperate search for the means necessary for their work. Their life was spent between the laboratory and their home and was made happy by the birth of two daughters Irène and Eve. In 1903 Pierre and Marie Curie received the Nobel Prize for Physics together with Henri Becquerel. In 1906 Pierre Curie died in a tragic accident. A little later Marie Curie was awarded the Nobel Prize for Chemistry.

After the death of her husband and for the whole of the rest

The Atom

of her life, which ended in 1934, Marie Curie remembered bitterly that a scientist such as her husband always had to struggle to obtain the means that were necessary for his work. In her biography of Pierre Curie, she writes:

'For the extraordinary gift of themselves and for the magnificent services which they supply to humanity what is the reward which our society offers to scientists? Are the servants of thought supplied with the apparatus for their work which they need? ... Our society dominated by a violent desire for luxury and greatness doesn't understand the value of science. It does not realize that science is part of our most precious moral heritage and it does not sufficiently understand that science is at the root of all progress which makes human life easier and which diminishes suffering. Neither public funds nor private generosity give to science and to the scientist the help and the support which are indispensable for fully efficient work.' These are words from the heart which preserve even today their full value.

In the course of their work Marie and Pierre Curie realized that some pitch-blendes (uranium minerals) which came from Joachimstal in Bohemia had a radioactive power which was greater than that of uranium. They thought then that the radioactivity of the Bohemian pitch-blendes came, not from the element uranium, but from another unknown body which was radioactive and which existed in minute amounts amongst uranium minerals. After a prolonged and intensive search which demanded many tons of material they discovered in 1898 the first new radioactive substance; they called it polonium in honour of the home country of Marie Curie. A few months later they succeeded in separating another element which had a radioactive power several million times stronger than that of uranium; this element was given the name of radium.

The discovery of radioactivity gave a strong impulse to the systematic study of uranium minerals, a study which very soon led to the discovery of nine different radioactive substances. Debierne discovered actinium, Boltwood ionium, Hoffman and Strauss radium D, etc.

As Soddy has noted, the major contribution made by Marie Curie to the progress of science was perhaps not the discovery of radium but the following principle which guided her research

—a principle which she quickly accepted and which she upheld despite some apparent exceptions. Radioactivity is a property of the atom which cannot be in any way influenced by any agent then known (exceedingly high temperature, exceedingly high pressures, light radiation, electric and magnetic fields etc.) and which does not change even when radioactive elements are chemically combined with other elements.

ALPHA RAYS, BETA RAYS AND GAMMA RAYS

The work went on. New experiments were devised and very delicate measurements were made. Chemists and physicists threw themselves into the study of the new phenomenon. The first subject studied was the nature of the radiation given off by radioactive substances. What type of radiation is it? Is it a corpuscular radiation—that is, made up of a flow of particles, or is it electromagnetic radiation like light and X-rays?

Between 1889 and 1900 Lord Rutherford realized that the rays emitted by radioactive substances are of three different types. He reached this conclusion by observing the different absorptions of the radiation by different materials. Not all rays are equally absorbed when they cross matter. The thickness of a given material which radiation can cross without being absorbed is a measure of its penetrating power.

Let us see for example what is the penetrating power of two types of ray which are known to everybody, light rays and X-rays. Light comes from the sun and reaches us after having crossed the whole atmosphere. This shows that a thickness of some hundreds of kilometres of air is not able to absorb much light but in general light cannot cross solid substances, even if these are not very thick. (We make an exception of the few substances like glass, which do not absorb light and which because of this are called transparent.) In particular, light rays cannot cross the human body and it is for this reason that we cannot see our lungs, our heart and our bones.

X-rays on the other hand, can cross the muscular parts of our body and are effectively absorbed only by the bones. If one sends a beam of X-rays on to a person and places behind him a photographic plate the X-rays cross the body and make an impression on the plate. The bones on the other hand absorb part of the rays and the corresponding parts of the plate receive a smaller impression (Plate 1). One can therefore say that X-rays have a penetrating power which is greater than that of light rays. Lord Rutherford found that the different radioactive radiations cannot all cross the same thickness of material. He made experiments with plates of aluminium and discovered that the amount of radiation transmitted depended on the thickness. One group of radiations was absorbed in a thin sheet only 0·01 mm. thick; the other radiations on the other hand could cross plates which were thicker than these. However, when the thickness of aluminium reached about 1 mm. another group of radiations became increasingly absorbed. The rays which crossed this last plate and which were therefore the most penetrating were absorbed only in a layer of aluminium about 10 centimetres thick.

The three groups of radiation distinguished by penetrating power were called respectively alpha rays, beta rays and gamma rays. One can therefore say that beta rays are approximately a hundred times more penetrating than alpha rays and that gamma rays are from ten to a hundred times more penetrating than beta rays.

Apart from their different penetrating powers alpha rays and beta rays can also be distinguished by their different behaviour in an electric field. Let us take a block of lead (Fig. 5) in which we make a small hole and inside this hole let us place a grain of radioactive substance. The block of lead then absorbs all the radiation emitted by the radioactive material except that which comes out from the opening of the hole in the form of a thin beam and in a well-defined direction.

Let us put this block between two parallel metal plates which are connected to the terminals of a battery so that one is positively charged and the other negatively charged. We then find that the beam becomes divided in three parts; one is deflected towards the positive plate, one goes forward in a straight line and the third is deflected towards the negative plate. These deflections can be revealed either by making impressions on a photographic plate or by another method (which we will explain later) which allows one to reveal the whole path of the rays.

The part of the beam which is attracted by the positive plate

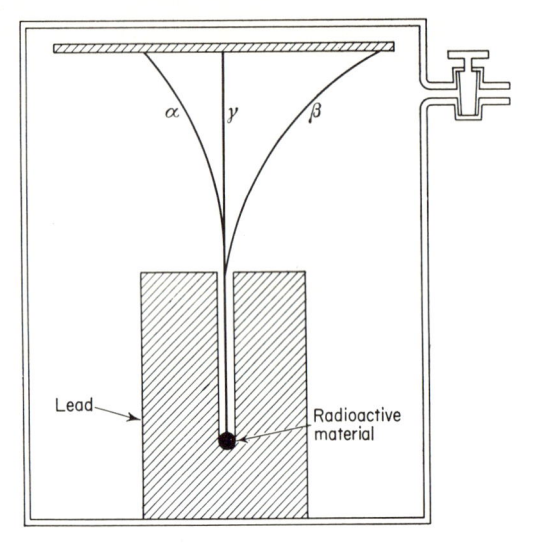

Fig. 5. Path of alpha, beta and gamma rays in an electric field.

should clearly be made up of negative charges and that which goes towards the negative plate must carry positive charges. The positive rays are those with the minimum penetrating power, i.e. the alpha radiation; the negative rays are those of intermediate penetrating power, the beta radiation and those which are not affected, and which do not therefore carry electric charge make up the gamma radiation.

THE BETA AND GAMMA RAYS

It was easier to determine the nature of beta and gamma rays than that of alpha rays.

In 1899 Henri Becquerel and others showed that beta rays are electrons emitted at very high velocity from radioactive substances. In some substances these electrons are emitted at a velocity which is very near the velocity of light (300,000 kilometres a second).

The fact that one part of the radiation (the gamma rays) does not undergo any deviation in an electric field or in a magnetic field shows that it is made up of particles which are electrically neutral and that it consists of electromagnetic radiation similar

The Atom

to light and X-rays. This was first recognized in 1900 by the Frenchman Paul Villard (1860–1933).

It is impossible to explain electromagnetic radiation in a few words. We will therefore be content to note only something about the way they travel.

Electromagnetic rays do not consist of the vibrations of material particles and therefore do not correspond to any type of wave propagation with which we are familiar. However to explain some of the properties it is convenient to take an example which, if it is not new to many, is always that which best represents the phenomenon of the propagation of waves irradiating from a point. Let us pause to consider what happens to the surface of a pond when a stone is thrown into it.

A series of circular waves are formed at the point where the stone hits the surface. These circular waves grow and move away

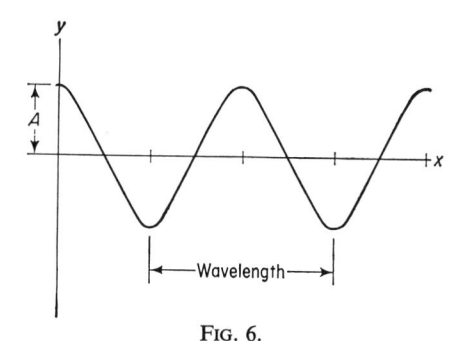

Fig. 6.

from the centre at which they are formed. The number of waves which are formed in a second can be found by counting the waves which break against the banks of the pond. One realizes intuitively that this number, which is the frequency of the waves, depends on the velocity with which the waves move and on the distance which lies between one wave crest and the next; this distance is called the wavelength (Fig. 6). 'The greater the velocity of the wave and the smaller the wavelength, the greater the frequency.'

Electromagnetic waves are more complicated because they travel throughout space and are not restricted to a plane as on the surface of a pond; they are therefore not circular but

The Nature of Matter

spherical; but a law analogous to that which we have just discussed is effectively valid. In space all electromagnetic waves travel with the same velocity—300,000 kilometres a second; thus for them the frequency depends only on the wavelength.

The electromagnetic radiations of longer wavelength and thus of lower frequency are those which are used in radio-telegraphy (Fig. 7). Moving towards shorter wavelengths one

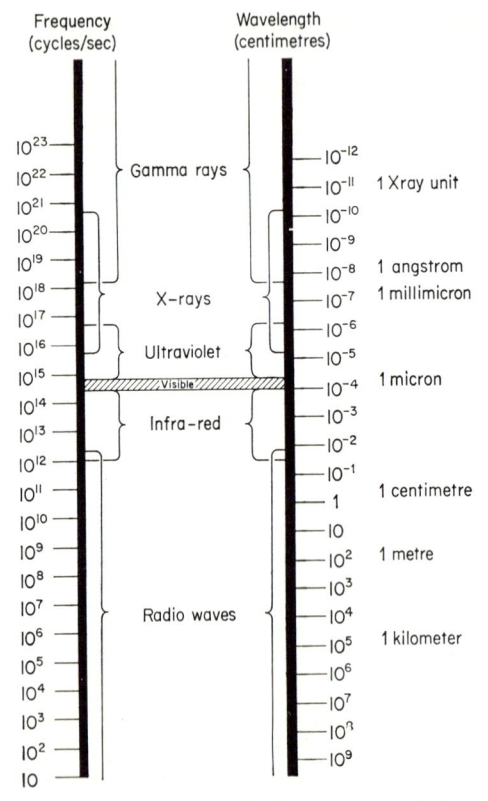

FIG. 7. Wavelengths of electromagnetic radiations.

finds infra-red radiation and then visible radiation. The latter is confined between two well-defined frequency limits. Red light which is 4.3×10^{14} waves per second forms the lower limit; the upper limit is given by the frequency of violet light which is 7.5×10^{14}. Radiations with frequency higher than this are no longer visible. They make up in order of increasing frequency X-rays and gamma rays.

Gamma rays then are electromagnetic waves with a wavelength very much smaller than that of X-rays. Because they are of shorter wavelength X-rays are more penetrating than light rays. They are able to cross obstacles which stop light. Because of their very much smaller wavelength gamma rays are very much more penetrating than X-rays. They are able to cross several thicknesses of lead. For example, the gamma rays emitted by radium can be detected after they have crossed thirty centimetres of lead.

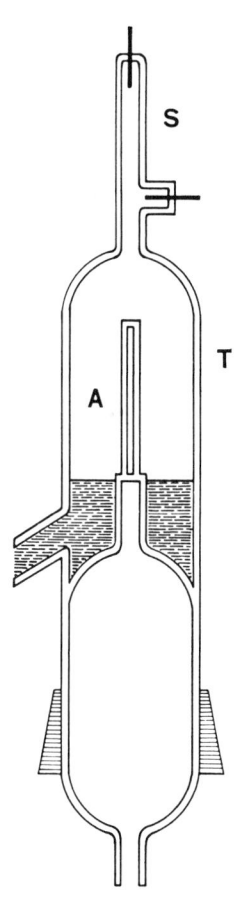

ALPHA RAYS

The research into the nature of alpha rays was very much more laborious, and was ended only in 1909 by Rutherford, who had first discovered them. He, working with various collaborators, made this radiation his special field of research.

The first investigations on alpha particles were able to establish only certain additional properties beyond the penetrating power which had already been measured by Rutherford. They are material particles very much heavier than beta particles and carry a positive electric charge. The velocity at which they are emitted varies from about 12,000 to 20,000 kilometres per second.

The nature of these particles was definitively established only much later as a result of an experiment made by Lord Rutherford in collaboration with the physicist Royds. They made use of an exceedingly simple apparatus. A small tube of glass (A) (Fig. 8) with walls sufficiently thin to allow passage of alpha particles was put in a very much bigger tube

FIG. 8. Rutherford and Royds' apparatus.

(T) also made of glass. The walls of this tube absorbed the alpha particles and allowed the beta and gamma radiation to pass through. Radium was put in the small tube and, using a pneumatic pump, a vacuum was made in the large tube. Two days after the radium had been introduced into the small tube (A) it was found that the outer tube (T) contained gas.

In fact, as a result of observations that had been made by other physicists, it had already been foreseen that this would happen. Rutherford and Royds had therefore arranged that another small tube (S) communicated with the tube (T), and using the pump they drove the gas into this small tube. In this way they were able to obtain a concentration high enough for spectroscopic study. They were able to establish that the gas collected in the tube (T) was helium, i.e. the element which occupies the second place in Mendeleev's table. (Atomic no. 2, atomic weight 4.)

The experiment was interpreted as follows. The alpha particles were 'fired' from the radium through the thin walls of the small tube (A) and then struck violently against the thicker walls of the tube (T) penetrating for a very small distance into the glass. In this way they lost their electric charge and gave rise to the formation of neutral helium atoms which were re-expelled from the glass into the tube (T) in which the vacuum had been established. The fact that the alpha particles, when they lose their positive charge, that is when they acquire electrons, change themselves into helium atoms show that each of them is an atom of helium which has lost electrons.

Later on we will discuss radioactivity in greater detail. We will now turn to the problem of the constitution of the atom.

RUTHERFORD'S COLLISION EXPERIMENTS

Rutherford had discovered the existence of alpha particles in the radiation emitted by radioactive substances and he had found out what these particles were. It was he again who had the brilliant idea of using them as a new weapon for the study of the constitution of the atom, i.e. for the study of the arrangement of the positive and negative electric charges in the atom.

Alpha particles are emitted by radioactive substances at extremely high velocity—tens of millions of kilometres per

The Atom

second. When these very fast projectiles (spontaneously supplied from radioactive substances) traverse matter they are deflected by their collision with atoms, for the size and mass of each one of them is not very different from the size and mass of any atom. The deflection suffered by an alpha particle in a collision with an atom will depend on the structure of the atom; that is, it will depend upon the way in which the mass is distributed in the atom; and since an alpha particle has a positive electric charge it will depend also on the way in which positive and negative charges are arranged in the atom. It was the brilliant idea of Rutherford to study these deflections and from them to deduce information on the constitution of atoms.

To do this many experiments using fine beams of alpha particles emitted from small quantities of radioactive substances were made by Rutherford himself and by his collaborators working under his direction. These beams were directed on to thin layers of metal and the emerging alpha particles were detected on photographic plates.

The results of these experiments were strange and unexpected. It was found that most of the particles were hardly deviated at all in passing through the metal layer; only very few were strongly deflected or repelled backwards (Fig. 9). For example, if an experiment is made with a thin layer of platinum we find that only one alpha particle in eight thousand is deflected through an angle greater than a right angle. How could this result be explained?

The fact that the greater part of the alpha particles passed almost undisturbed through the metal layer showed that atoms consist essentially of empty space, in conflict with the old concept of the atom as a solid piece of matter.

Furthermore the large deviation suffered by some of the alpha particles could not have been due to collisions against the electrons contained in the atom, because each electron has a mass which is some thousands of times smaller than the mass of the alpha particles. Each one of these large deviations should therefore be due to the influence on the alpha particles of that part of an atom which contains almost all the mass and all the positive electric charge.

As an alpha particle has a positive electric charge the collision against the part of the atom which is positively charged will not be a mechanical collision but will be an electric collision; this means that the positive charge of the atom will repel with electric force an approaching alpha particle; and the smaller the distance of approach the greater will be the force. But the repulsive force exerted on the alpha particle by the positive part of the atom—and as a result the path followed by the particle

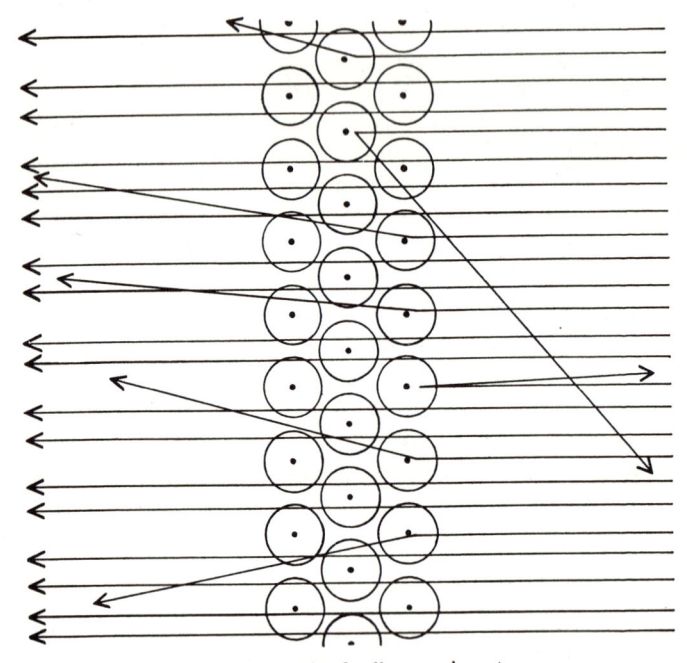

FIG. 9. Rutherford's experiment.

after the collision—will vary according to the distribution in the atom of the positive electric charge.

But in order to explain the deflections found in these experiments Rutherford was obliged to conclude that there exists in the atom a strong electric field. This field must be much stronger than that which would exist in an atom consisting of a homogeneous sphere of positive electricity (Fig. 10), as in the model proposed by Thomson.

Rutherford therefore proposed in 1911 a model of the atom in which there existed the strong electric field required to explain the results of these experiments. He supposed that the positive charge of the atom was concentrated in a small zone at the centre (a little later this small central zone was given the name of 'nucleus'). The dimensions of the nucleus are extremely small compared to the dimensions of the atom; and in the nucleus is concentrated almost all of the mass of the atom.

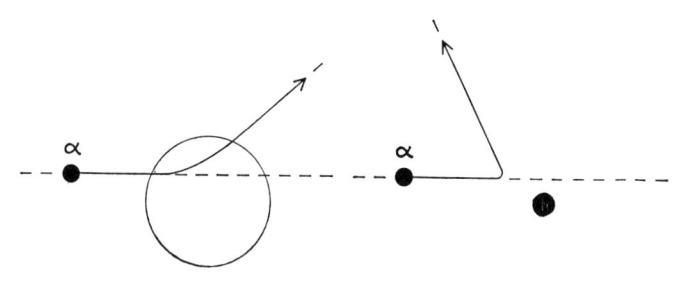

FIG. 10. Deflection of an alpha particle in the case of Thomson's atomic model and in the case of Rutherford's model.

With this hypothesis Rutherford calculated how the deflections of the alpha particles would be distributed through various angles. This calculation was shown to be in full agreement with the results of a beautiful series of experiments carried out on the suggestion of Rutherford by Geiger and Marsden in 1913. The conclusions drawn by Rutherford on his nuclear study of the atom were confirmed point by point.

THE ATOMIC NUCLEUS

Experiments therefore fully confirmed the validity of Rutherford's ideas. The atoms of all the elements are made up according to a single general model. Every atom contains an exceedingly small nucleus, which carries all the positive charge and almost all the mass of the atom.

Since the atom considered as a whole is electrically neutral it must contain in addition to the nucleus a number of electrons such that the combined negative electric charge of the electrons is equal to the positive electric charge carried by the nucleus. These electrons surround the nucleus and 'protect' it from everything which arrives from outside except from particles with very high energy.

In order to express the value of the positive electric charge of the nucleus we must fix a unit suitable for measurement. Now as the electric charge of an electron is the smallest quantity of electricity that is known, it is taken in atomic physics to be, e, the unit of charge. Since on the other hand the charge of the electron is, as we know, negative, it is indicated as -1; then $+1$ indicates the unit positive electric charge which is equal, apart from sign, to that of the electron.

What then is the electric charge of a nucleus expressed in these units?

The experiments of Geiger and Marsden had already supplied a rough estimate of the value of this charge; and in the same year (1913) Van der Broeck suggested that the electric charge of the nucleus of an element (expressed in the units defined above) has a value equal to that which is called the atomic number Z of the element. This number refers to the order in which the element appears in Mendeleev's table in which, as we know, the elements are arranged in order of increasing atomic weight. The nucleus of an atom of hydrogen would then have a positive charge whose value is one. The nucleus of helium would have a charge two ... the nucleus of gold 79 ... the nucleus of uranium 92 (that is equal respectively to 1, 2, 79, 92 times the value of the electric charge of an electron).

The experiments of Moseley on X-ray spectra were the first to fully confirm this idea. Its importance and wide range of validity were very quickly recognized.

Thus the atomic number Z of the element which, up to that time, had simply been the number indicating the position of the element in the periodic table acquired a real physical significance. It represents in units of e the value of the positive electric charge of the atomic nucleus of the element. Bohr wrote 'One can say that this interpretation of the atomic number constituted an important step towards the solution of a problem which was for long one of the boldest dreams of scientists; to construct an explanation of the laws of nature based on simple considerations of numbers.'

I have shown that the atomic nucleus occupies only an extremely small part at the centre of the atom. In fact the collision experiments of Rutherford were to show that the radius of the nucleus is (for the heavier elements) 10^{-12} centimetres (we

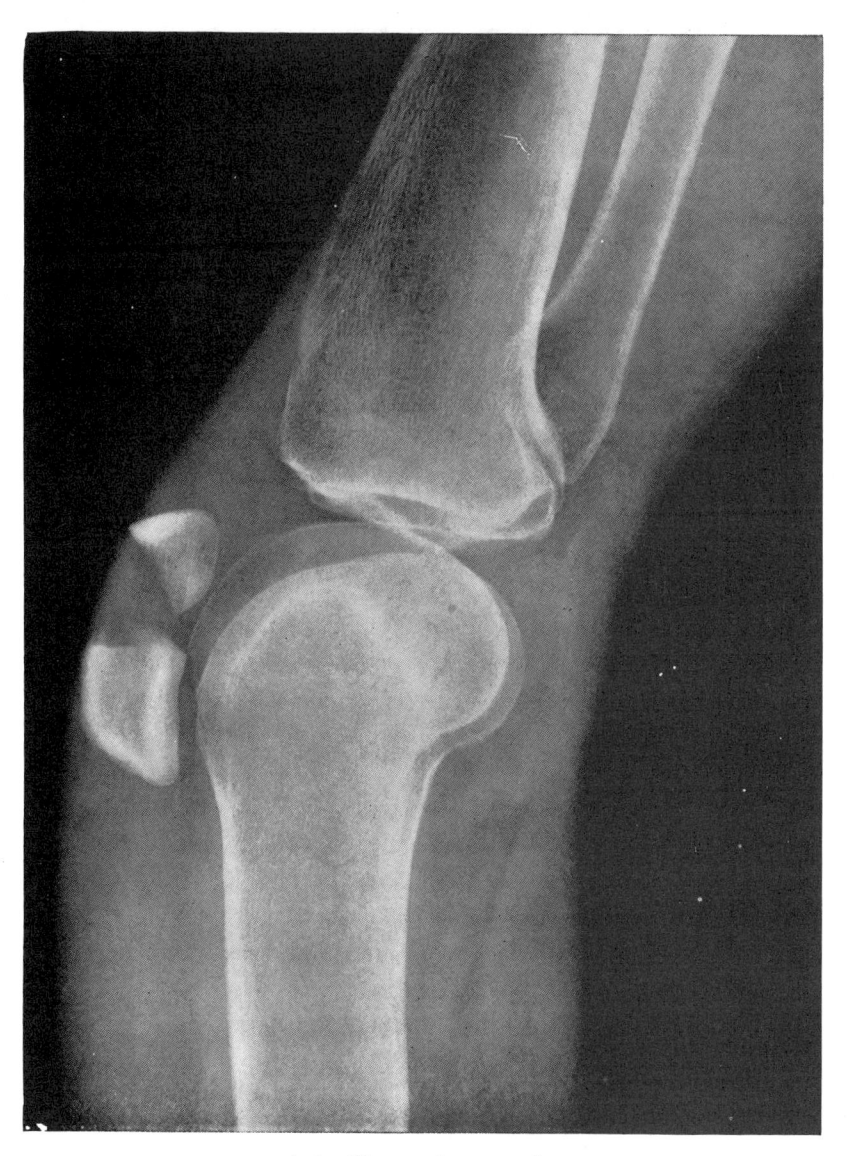

1 An X-ray photograph

2a Tracks of alpha particles in a Wilson chamber with magnetic field

2b Tracks of beta rays in a Wilson chamber with magnetic field

remember that the radius of an atom is about 10^{-8} centimetres). This means that whilst an atom has a diameter of the order of a hundredth millionth part of a centimetre a nucleus occupies a region whose diameter is of the order of some millionths of a millionth of a centimetre.

In order to get an idea of the relative dimensions of the atom and of its nucleus we can imagine expanding an atom until it becomes a sphere with the diameter of a mile. The nucleus would then appear at the centre of this sphere as a ball of a few inches diameter.

We therefore conclude that an atom (just as the planetary system) is a region of space which is essentially empty.

THE ATOMIC MODEL OF RUTHERFORD

This then was the model of the atom which Rutherford proposed in 1911. It formed the basis of all the developments in atomic physics up till 1925. Even today it is considered as a good approximation of the atom and remains of great importance not only from a historical or didactic point of view but also because it provides the basis for a vocabulary which is particularly convenient and expressive.

According to this famous model the atom behaves as a miniature solar system. At the centre is the nucleus (the sun) of this system in which is concentrated almost all of the mass and all of the positive charge of the atom; the electrons revolve around the nucleus in eliptical orbits just as planets revolve around the sun. Indeed, the electrons which have a negative electric charge are attracted to the positive nucleus by an electric force which, like the gravitational force which binds the planets to the sun, is proportional to the square of the distance.

The atom, like the planetary system, is mostly empty space and just as almost all of the mass of the planetary system is concentrated in the sun so almost all of the mass of the atom is concentrated in the atomic nucleus. But the sun is very much larger than the planets, which all have different dimensions; on the other hand, the atomic nucleus has about the same dimensions as an electron and the planetary electrons are all of the same size.

With Rutherford's atomic model atomic theory became an

The Nature of Matter

imposing construction and the problem of the structure of the atom appeared practically solved.

The atoms of each of the 92 elements that exist in nature are made up of a central nucleus (in which is concentrated almost all the mass and all the positive charge of the atom) around which revolves a certain number of electrons. The greater the electric charge of the nucleus the greater is this number of electrons; it increases by one when we move, in Mendeleev's periodic table, from one element to the next.

The hydrogen atom which is the lightest element is made up of a nucleus (which takes the name of proton from the Greek word protos meaning first) with charge plus one around which revolves a single electron at the distance (in normal conditions) of about one-hundredth millionth of a centimetre. The simplest element after hydrogen is helium. The atom is made up of a central nucleus of charge plus two around which revolve two electrons. The atomic weight of helium is four. This means that the atom of this element is four times heavier than the atom of hydrogen which has been chosen as the unit for measurement.

Next comes lithium whose atom is made up of a nucleus of charge plus three, around which three electrons revolve. And so on, until we reach uranium the heaviest element existing in nature. Its atom is made up of a nucleus of charge plus 92, surrounded by 92 planetary electrons. In general, the atom of an element whose atomic number is Z is made up of a nucleus of charge plus Z around which revolves Z electrons (always assuming that our unit of charge is the charge of an electron).

Thus in the atomic theory of Rutherford the fundamental characteristic of an atom is the value of the electric charge of its nucleus, that is its atomic number. It is this which fixes the number of planetary electrons and as a result determines the chemical properties of the particular atom.

Before going farther I would like to be rather more precise about one point. I have just said that even today it is believed that Rutherford's hypothesis on the structure of the atom accounts well for the real world; but I have not said that Rutherford's atom is a good representation of reality. This is because Rutherford's representation, like other representations which we will meet later, are only models which are useful in helping us understand the behaviour of atoms. When I say 'the atom consists of a minute planetary system' one should not think that atoms are really made in this way but rather that phenomena of atomic origin which we observe behave *as if* the atom consisted of a miniature planetary system.

ATOMIC PHYSICS AND NUCLEAR PHYSICS

According to Rutherford's atomic model the chemical and physical properties of the atom depend on the number and on the movement of the planetary electrons and the reason for the great stability of normal elements (i.e. those which are not radioactive) is due just to the fact that ordinary physical and chemical reactions change only the arrangement of the planetary electrons of the atom without touching the nucleus. On the other hand, the radioactive properties of a radioactive atom are not affected by chemical changes, by X-rays or by any physical agents (temperature, pressure and electric discharge, etc.) except collision with particles or with very penetrating radiation. This shows that the radioactive properties of an atom derive from its nucleus and are not influenced by the planetary electrons.

One can therefore distinguish between 'atomic physics' and 'nuclear physics'. The object of atomic physics is the investigation of the behaviour of the planetary electrons of the atoms. In this investigation the nucleus of the atom appears only as a single particle of negligible size which is characterized only by the value of its mass and of its positive electric charge (that is of its atomic number). Atomic physics therefore ignore completely the structure of the nucleus; this is the subject studied in nuclear physics.

It is natural to consider first atomic physics; later we will enter into the smaller world of the nucleus.

CHAPTER THREE

ATOMIC PHYSICS

IONIZATION

We have seen that an atom is normally neutral; that is in normal conditions the positive electric charge of its nucleus is balanced by the negative electric charges contributed by the planetary electrons.

In some circumstances however the atom can lose one or other of its electrons; and because an electron carries a negative charge the atom after this loss is no longer electrically neutral but is positively charged. An atom which is mutilated in this way is called a positive ion; if it has lost a single electron it is a monovalent positive ion; if it has lost two it is a bivalent positive ion etc.

If an electron is to be torn away from the atom it must obtain the energy which is necessary to overcome the force which binds it to the atom. This energy can only be supplied to it from outside by radiation which falls on it (X-rays, ultra-violet rays light . . .) or by a very fast particle which collides with it. These collisions can be produced either by a suitable rise in temperature or by the passage of an electric discharge.

If the energy supplied by radiation or by a particle is high enough it can extract from an atom not only one electron but two, three . . . up to all of the satellite electrons. One then says that the atom is completely ionized. Although these conditions are rarely achieved on earth they are common enough in the stars in which there are temperatures of several millions of degrees and in which there are X-rays and very fast particles. In fact in the stars almost all the atoms are completely ionized; they are reduced to single nuclei deprived of all their satellite electrons.

An atom can, for a time, as we shall soon see, acquire a few additional electrons; it is then negatively charged; it is a negative ion. It will be a negative ion which is monovalent or bivalent or trivalent, etc., according to whether there are one, two or three . . . additional electrons.

And now that we understand the phenomenon of the ionization of atoms we can explain the origin of the cathode rays and the canal rays which are observed in a Crookes' tube. We will therefore now return to Rutherford's atomic model and we shall see what are the laws which reign in the microscopic planetary system.

When an electric discharge occurs in the tube planetary electrons are torn from some of the atoms of the cathode and from many of the atoms of the gas. In the inside of the tube there are three types of free particles. Atoms of gas which are still intact, free electrons and positive ions (i.e. those atoms which have lost electrons).

The electrons which have a negative charge are strongly repelled by the cathode which is itself negatively charged and strike violently against the walls of the tube opposite the cathode. These are the cathode rays. In fact experiment shows that they are made up of electrons.

Positive ions, on the other hand, are attracted to the cathode and rush towards it so fast that if it is perforated they pass through the holes and strike against the walls which lie behind the cathode. These are the canal rays and in fact, experiment shows that they are made of positively charged particles. The mass of these particles depends on the gas which is contained in the tube.

So we see how our understanding of the phenomenon of ionization allows us to explain what happens in a Crookes' tube when an electric current flows between the anode and the cathode.

NEW DIFFICULTIES

We will now return to the model proposed by Rutherford. According to this model the atom should be regarded as an exceedingly small planetary system in which the nucleus takes the place of the sun and the electrons, which are bound by electrostatic attraction, revolve around the nucleus as the planets revolve around the sun.

Both in the atom and the planetary system the central body has a mass which is very much greater than the mass of the bodies which revolve around it. In both the force which acts between the central body and those surrounding it, although of a different type in the two cases, is inversely proportional to the square of the distance between the bodies.

There is, however, an essential difference between the atom and the planetary system; the planets belonging to the atom, that is the electrons, have an electric charge. Now according to the fundamental law of electromagnetic theory a body which is charged with electricity and which does not move steadily in a straight line (that is a body which undergoes an acceleration) should continuously radiate energy. Thus every electron in the atom as it turns around the nucleus should lose energy by radiation. The electrons should continuously approach the nucleus. They should travel on a spiral path and they should end by falling into the nucleus. The atom would cease to exist as such after no more than a hundredth of a millionth of a second. We see therefore by applying to Rutherford's atom the classical laws of electromagnetic theory we reach the absurd conclusion that atoms are not stable systems; but the very fact that matter still exists tells us that atoms (apart from radioactive atoms) must be stable.

There exists another difficulty. According again to classical theory an electron which revolves around the nucleus should radiate energy in the form of electromagnetic radiation (i.e. light, X-rays ...). The fundamental frequency of this radiation should be the rotation frequency of the electron; but we have seen that according to the laws of electromagnetism the electron should radiate energy and should approach closer and closer to the nucleus (until it falls into the nucleus and neutralizes it). Its orbit should become smaller and smaller; thus the frequency of its motion should continuously vary and as a result the frequency of the light emitted should also vary continuously. Now every body is made up of an enormous number of atoms and all the possible ages in the atomic life span should be represented. Thus every body that is able to emit electromagnetic radiation should emit all possible frequencies. This conclusion is in complete contradiction with experiment. We shall now see why.

Atomic Physics

Let us suppose that we allow a ray of sunlight to fall on one side of a prism and that we collect on a screen the ray which emerges from the prism. As it crosses the prism the ray will undergo dispersion so that we shall see on the screen not a single point of light but a strip of light containing a series of colours. This coloured strip is called a spectrum. In order to observe the spectrum of the light of a given source one uses an instrument which consists essentially of a suitably arranged prism and lenses. This instrument which is called a spectroscope was invented in 1859 by Bunsen and Kirchhoff.

The study of the spectrum of the light from a given source implies therefore the separation of light of different colours, that is the separation of radiation of different frequencies.

Every source of light produces a spectrum whose nature depends on the source employed. If in particular one studies the spectrum of an incandescent gas (for example neon or argon) one sees that it is made up of a number of bright lines of different colours on a dark background. Experiments have shown that there is a characteristic spectrum for each element. For example when one observes the spectrum of neon one always sees the same lines occupying the same positions (that is lines always of the same frequencies), both when the neon is in a pure state and when it is mixed with other gases. Experiment therefore shows that the atoms of a gas do not emit radiations of all possible frequencies as the application of electromagnetic laws to Rutherford's atomic model would lead us to expect; they emit only certain radiations of well-defined frequencies.

Rutherford's atomic model gives therefore two results which are in complete disagreement with experiment if the ordinary laws of mechanics and electromagnetism are applied to it.

Poincaré, noting the failure of the classical concepts, prophesied that 'therein lies one of the mcst important secrets of nature'.

QUANTA

How were these difficulties to be overcome? Was it necessary to give up Rutherford's atomic model even though it accounted so elegantly for the periodic table of elements?

It was not however the first time that physicists came across contradictions and difficulties in applying to the atom the ordinary laws of mechanics and of electromagnetism. They began to be convinced that these laws were no longer valid when one moved from the macroscopic world to the microscopic world of the atom. It is not altogether surprising when one thinks that there is no *a priori* reason why laws which are valid in the macroscopic world should be transferred without any modification to systems which are thousands of millions of times smaller.

Some attempts to substitute for these laws new laws which would account better for the results of experiments had already met with success. The new laws were not based on an extension of the ordinary ways of thinking of classical physics; they represented an empirical modification of classical ideas which was made in order to put theoretical deductions in harmony with the results of experiment.

The first of these attempts had already been made in 1900 by the German physicist, Max Planck. It was extended in 1905 by another German physicist, Albert Einstein. It is not possible now to dwell in detail on these studies which opened the way to the *Quantum theory*. I will only say enough to explain what were the new concepts which allowed one to establish new laws which were proved to be valid in the atomic world.

The so-called *black body* is a body which is able to absorb all the radiation which falls on it (black smoke nearly satisfies this condition). The application of the ordinary laws of mechanics and electromagnetism in the theoretical study of the radiation emitted by a black body led to a formula which was in complete contradiction with the results of experiment. Several fruitless attempts were made to overcome this contradiction before, towards the end of 1899, Planck discovered a surprising fact. The results of theoretical calculations were in full agreement with the results of experimental measurements if one made the following hypothesis (expressed in modern language): when an atom emits energy in the form of radiation of frequency ν this energy is not emitted continuously but only in amounts which are multiples of an elementary quantity ϵ. To this quantity ϵ Planck gave the name of *quantum of energy*.

Now we know that every electromagnetic ray is characterized by its frequency. What then is the frequency of electromagnetic rays which contain one quantum of a certain energy?

Atomic Physics

Planck discovered in 1900 the relation which links the energy of a quantum to the frequency of the corresponding electromagnetic waves: every quantum has an energy which is proportional to its frequency. In simple words this means: *the greater the frequency v, of the emitted radiation the greater is the amount of energy ϵ which makes up the quantum*; the frequency of the radiation and the energy of the corresponding quantum are related by the simple equation $\epsilon = hv$ where h is an exceedingly small constant ($h = 6 \cdot 610 \times 10^{-27}$ erg sec). It was then shown that this *Planck's constant* is a universal constant. In later work this constant became more and more important.

Now this hypothesis of Planck's that energy is emitted in discrete packets was extremely bold because, according to the fundamental principles of classical physics, all actions can be continuously varied. But only this hypothesis could account for experimental results and Planck, after long hesitation, put it forward before the Physical Society of Berlin on December 14, 1900.

Thirty years later the physicist Bohr was to write, 'There is no doubt that there exists in the history of science very few events which in the brief period of a generation have had consequences as important as the discovery of Planck'.

And so was born the idea that classical laws of physics are not appropriate for the atom. Then only five years later Albert Einstein extended Planck's hypothesis in order to overcome another serious contradiction between the wave theory of light and certain experimental facts (concerning the photo-electric effect which we will consider later). Planck had shown that there was a discontinuity in the emission and absorption of radiation. Einstein showed that the discontinuity applied not only to emission and absorption but that radiant energy was localized in grains when it travelled through space (and was not distributed over a wave front as required by the wave theory of light). To these grains of energy Einstein gave the name of light quanta or photons.

The equation discovered by Einstein for the photo-electric effect is then based on a concept of radiation which is in complete disagreement with the fundamental postulates and conclusions of the electromagnetic theory of radiation. It was completely verified by experiment and became more and more widely applied.

It should be remembered that although the two hypotheses of Planck and of Einstein reinforce each other they are distinct. Planck's hypothesis of quanta referred to the manner in which energy is emitted and absorbed by the atoms which make up matter, while Einstein's hypothesis of light quanta refers to the manner in which energy travels through space.

Planck's theory has acquired the name of the *Quantum Theory of Radiation* whilst the expression *Classical Theory of Radiation* refers to the theory before Planck's hypothesis.

The quantum hypothesis was therefore confirmed by the theory of black body radiation and by Einstein's theory of the photo-electric effect. But it was very soon shown that it was valid in the very many different fields and the value of the constant h obtained from the study of many different phenomena are in remarkable agreement. 'The brilliant and original idea of Planck', wrote de Broglie, 'was by about 1913 established by very many experimental facts. Bohr's atomic theory arrived at that moment to give it new spectacular confirmation showing how far the very structure of matter is determined by the existence of quanta.' This quantum theory, this new concept of light being composed of particles has become more and more firmly established; there have however been enormous difficulties in reconciling it with the wave theory of light. Only the wave theory can explain the phenomena of interference and diffraction. We shall see later how this dualism has been accommodated and how in a certain sense a reconciliation has been achieved.

The success of the quantum theory does not prove that classical theories are false. It proves merely that classical theories which were deduced only from observations of the behaviour of macroscopic bodies, though valid in this field, are inadequate to explain the microscopic structure of the universe.

BOHR'S HYPOTHESES

When the classical laws of electromagnetism were applied to Rutherford's atomic model there arose the difficulty to which I referred earlier, that is the impossibility of accounting both

Atomic Physics

for the stability of atoms and for the existence of lines in the spectra of gases; so that it had for long been supposed that the laws which applied to the inside of the atom were different from the classical laws.

In 1912 the Danish physicist Niels Bohr, noting that none of the attempts which had so far been made had succeeded in explaining the stability of atoms, became convinced that it was necessary to introduce a new principle, quite foreign to the physics then existing. Instead of choosing in an arbitrary way the necessary modifications of the fundamentals of physics Bohr based himself on Planck's and Einstein's theory of the quantum nature of light; and in 1913 he put forward his quantum theory of the atom to a meeting of the Solvay Council. His proposal dealt in particular with the hydrogen atom.

Bohr accepted Rutherford's model of the atom according to which an atom of hydrogen consists of a central nucleus around which a single electron revolves. He added however new hypotheses.

According to the quantum theory of light every time that a system emits (or absorbs) light of frequency ν the amount of energy emitted (or absorbed) is proportional to the frequency (and is given as I have said by $h\nu$ where h is a constant). Thus the fact that an atom of a gas does not emit radiation of all frequencies but emits only radiation of certain well-defined frequencies implies as a consequence that the energy of an atom cannot have an arbitrary value but only certain discrete values. It is said that an atom can exist only in certain defined states, *energy levels* or *quantum states*.

As long as an atom stays in one of these levels it emits no energy; but when it moves from a state of higher energy to one of lower energy it emits a quantum, ϵ, in the form of electromagnetic radiation of frequency ν (related to ϵ by $\epsilon = h\nu$); and ϵ is thus equal to the difference in energy between the two levels. On the other hand when an atom is struck by radiation of frequency ν and absorbs a quantum of energy ϵ ($= h\nu$) it moves from a lower energy level to a higher level, such that the difference in energy between the two levels is equal to ϵ.

Bohr discovered how to apply these ideas to a hydrogen atom treated according to Rutherford's model. He did this by making four hypotheses:

(1) The electron cannot revolve around the nucleus in any arbitrary circular orbit (that is on a circle of arbitrary radius) but is allowed to move only on a certain number of privileged orbits each of which belonged to a certain energy;

(2) As long as it moves in one of these orbits the electron does not irradiate energy;

(3) The electron can jump spontaneously from one of the permitted orbits to another permitted orbit of lower energy (that is from a larger orbit to a smaller orbit). In doing this it emits radiation containing an amount of energy equal to the difference in energy between the two orbits.

(4) When the electron jumps between the two orbits whose energy difference is ϵ the frequency of the emitted radiation was related to ϵ by $\epsilon = h\nu$.

This atomic theory of Bohr accounted for the stability of atoms and for the existence of lines in the spectrum. As the electron can exist only in a discrete number of orbits (we still limit our attention to the hydrogen atom) the atom can emit only certain fixed values of energy, i.e. only radiation of certain fixed frequencies. Each value corresponds to a jump between any two of the allowed orbits, the first of which having a higher energy than the second (that is the first orbit is larger than the second). The reason why we can see a large number of lines simultaneously in the hydrogen spectrum (or in the spectrum of other elements) is that the incandescent gas is made up of an enormous number of atoms. In each of these atoms the electron may be in any of the allowed orbits and may jump to any other allowed orbit of lower energy. The simultaneous production of spectrum lines has been described as a cumulative effect produced by an orchestra of atoms.

Among all the allowed orbits that of lowest energy is the smallest where the electron is closest to the nucleus and is bound to it as tightly as possible. In normal conditions the electron exists in this lowest orbit. One then says that the atom exists in the ground state because this is the most stable of the states. When the atom is in this ground state, that is when the electron revolves on the smallest possible orbit, the atom clearly cannot emit energy. Thus in order that an atom of hydrogen should emit radiation it is essential that its electron should not be in the innermost orbit; that is, it is essential that before

Atomic Physics

radiating, the atom should by some means such as a collision with a fast particle or by light, etc., be supplied with a certain amount of energy in order to make it pass from its ground state to a state of higher energy or equivalently that the electron should be displaced from the innermost orbit to an orbit of higher radius.

If, for example, the electron is in the second orbit it will not be in a stable state because sooner or later it will return to revolve in the first orbit, undergoing the process opposite to that which we have just considered. In the second orbit it possessed an energy greater than that which it now has in the first and in going from the second orbit to the first it emits the surplus of energy.

A similar thing happens if the electron is in the third, fourth or in a still higher orbit. It always finishes by jumping into the innermost orbit. If it jumps directly to the innermost orbit the value of the energy emitted will be the sum of the energies which would be emitted in a series of jumps from the fourth to the third, from the third to the second, from the second to the first. If, on the other hand, the transition occurs by a series of successive jumps the total energy emitted will be the same though it will be sub-divided into a number of quanta.

What I have said about the single electron of hydrogen could be repeated for all the electrons of other atoms. When one of these electrons exists in an orbit which is larger than its normal orbit its state is unstable and the atom is said to be excited.

A rather crude analogy to the assembly of electron orbits in an atom is provided by a circular staircase like that in a stadium. The highest energy orbits would correspond to the higher steps. Indeed the potential energy of a body placed on one of the higher steps is greater than that which it would have if it were on a lower step, just as the energy of electrons are greater when they are revolving in external orbits. We could put on the top step of the staircase a rubber ball which would represent the electron. The ball would be in an unstable condition and it would tend to fall down the staircase, jumping from step to step until it reached the bottom where it would stop.

The ball could make jumps of different sizes according to the difference in level between the step from which it started and the step on which it landed, but it can only make jumps through certain fixed differences in level because the differences in level between the various steps are fixed and cannot be arbitrarily varied. For example, the ball could jump through one foot only if there existed two steps whose difference in height was just one foot. In the same way, the electron of an excited atom can jump between various pairs of orbits which means that the energy emitted by the atom can have various values; but as all the orbits have well-defined energies the electron can jump only through certain fixed differences of level, and correspondingly the atom can emit only certain defined values of the energy. So a certain energy E can only be emitted by an atom if there exists two free orbits for which the difference in energy is E.

The theory which Bohr introduced for the hydrogen atom allowed him to calculate *a priori* the value of a constant which spectroscopists had already determined very precisely by experiment (the so-called Rydberg constant). The calculated and the experimental values were exactly the same. This was the first great success of Bohr's theory. The theory represents a remarkable combination of principles from classical theory with hypotheses which are a natural generalization of those which had been introduced by Planck and by Einstein, and which are radically different from classical theory.

THE CORRESPONDENCE PRINCIPLE

If one observes the spectrum of any element, for example that of hydrogen, one sees that the lines which are characteristic of the element do not all have the same intensity. Some are very much weaker than others. Although Bohr's theory allows one to calculate the frequency of spectral lines (that is, their position in the spectrum) it does not provide any means for calculating either their relative intensity or their state of polarization. From this point of view the classical theory of electromagnetism appears in a certain sense superior. The classical theory provides the means for determining all three of the characteristics of the lines, the frequency, intensity and polarization, even though the frequency calculated in this way turns out to be in disagreement with experiment (and as we have seen it was just for this reason that Bohr introduced his new hypotheses).

And so Bohr introduced his Correspondence Principle (despite the fundamental conflict between classical theory and quantum theory Bohr always tried to use for the construction of quantum theory concepts from classical theory appropriately modified). He started from the following consideration which arises from mathematical analysis. When a spectral line is produced by the transition of an electron from a very external orbit to another neighbouring external orbit (which implies a very small difference in energy) then the frequency of the line calculated from quantum theory tended to be the same as the frequency calculated by classical methods. The closer the distance between the neighbouring orbits and the more external the orbits the better is the agreement between the classical result and that derived with quantum mechanics. In these limiting conditions quantum theory coincides with classical theory.

Starting from this result which arises from both calculation and experiment Bohr tried to overcome the inadequacies of the theory which as we have said provided no way of calculating the polarization and the relative intensity of the spectral lines. He made the hypothesis that the intensity and the state of polarization of a line are related to the intensity and state of polarization of the corresponding lines which would have been emitted by the system according to classical theory and that this correspondence tends to become an identity in the limit of large quantum numbers.

In this sense there is a correspondence between classical physics and the physics of quanta and so Bohr gave to his hypothesis the name of 'Correspondence Principle'. Thus the new quantum physics is based in the limits on classical theory.

The introduction of this principle which had no theoretical justification may appear completely arbitrary. But it was empirically confirmed and has been of the greatest use in the further development of quantum theory because it provided in some cases a basis for arriving at quantitative results. Furthermore it was fully justified by the new quantum theory to which I will refer later.

SOMMERFELD'S CONDITIONS

It was in 1913 that Bohr put forward his hypothesis which he applied only to the circular orbits of the planetary electrons of the hydrogen atom. In 1914 Franck and Hertz carried out the first of a series of experiments on collisions between atoms and free electrons which directly confirmed the existence in any atom of different energy states with discrete values of energy; they demonstrated the validity of the formula which relates the frequency of the emitted radiation to the difference in energy between two quantum states. In 1916 the German physicist Arnold Sommerfeld extended to the elliptical orbits of the hydrogen atom the hypotheses which Bohr had made only for circular orbits; and he extended this hypothesis by applying to the motion of the electron not the classical mechanics of Newton but the relativistic mechanics of Einstein.

This led to a very important result. Sommerfeld found that the elliptical orbits which the electrons can follow around the nucleus are not equally spaced as are the steps on a staircase but are associated in groups which have been given the name of electron layers or shells. The orbits of the same layer have energies which differ by very little one from the other. It is as if a staircase was made of groups of very shallow steps and that each group was separated from the next by a much deeper step.

By convention the layers are called by the letters K, L, M, N ... moving from the innermost orbit outwards. The layers do not always contain the same number of orbits. The K layer contains two orbits, the L layer eight orbits, the M layer eighteen orbits, the N layer thirty-two orbits, etc.

In this way Sommerfeld's theory was able to explain a new complication which had appeared in the hydrogen spectrum when it was observed with a spectroscope with a very high resolving power—that was, that the lines are not simple but are in reality made up of several lines exceedingly close to each other.

One might think, as indeed it was thought, when Sommerfeld published his results that the theory of the atom refined in this way had already reached its final form. This conviction appeared at first to be justified by the result of the researches made by spectroscopists which also appeared to confirm the validity of Bohr's theory and the Correspondence Principle. But it was soon realized that the final form had not been reached. A more careful study showed that even Sommerfeld's theory which had applied relativistic ideas to Bohr's theory was unable to account quantitatively for the complexity of actual spectra; this complexity was revealed as the result of an even more detailed examination and of enormous effort in interpreting these experimental results. Sommerfeld's theory was also unable to explain certain anomalies in the magnetic properties of the atom.

PROBABILITY

According to Bohr's theory an atom emits energy when one of its primary electrons jumps from one orbit to another which is closer to the nucleus (i.e. to another which is associated with lower energy). But it is not possible to predict when a jump of a given energy will occur nor how long an electron will revolve in a given orbit.

As Bohr himself wrote: 'it was necessary to give up describing the behaviour of individual atoms in space and time according to the principle of causality and to imagine that nature could make amongst various possibilities a free choice which was not governed by any considerations other than probability.'

Bohr's theory then could not give information on individual processes which occurred in the inside of the atom. In 1917 Albert Einstein introduced coefficients of probability into the study of Bohr's atom.

He made the hypothesis that a planetary electron of an atom which revolves in a given orbit other than that of lowest energy has in a given interval of time, a certain probability, which is independent of external conditions to transfer spontaneously into a more interior orbit. Further he supposed that if an atom is struck by light whose frequency is equal to the frequency corresponding to the jump between two orbits that an electron may be transferred from a lower energy orbit to a more exterior orbit; and that the probability that this transfer takes place is the greater the greater the intensity of the light which strikes the atom.

Thus, for the first time in the mathematics of quantum theory probability factors were systematically introduced; the significance of these probability factors is much deeper than was at first realized.

THE HYPOTHESIS OF THE SPINNING ELECTRON

There are certain important details in the emission and absorption spectrum of atoms (certain phenomena such as ferromagnetism and paramagnetism) which at first could not be explained on Bohr's atomic model; but in 1925 the physicists Uhlenbeck and Goudsmit completed this theory with a new hypothesis thanks to which the difficulties mentioned above disappear. This hypothesis has been of fundamental importance in the further development of atomic physics.

Uhlenbeck and Goudsmit supposed that an electron not only revolves in an orbit around the nucleus but at the same time revolves around itself, just as the earth not only moves around the sun but also spins on its own axis. As a result of this spin (around an axis passing through its centre of gravity) the electron acquires properties just like those of a rapidly rotating top or gyroscope. It is well known that the axis of a gyroscope tends to remain in a well-defined direction.

As the electron is spinning and is electrically charged it follows also that it behaves like a little magnet with its axis parallel to the axis of spin. This means that if the electron is placed in a magnetic field its axis will orientate itself to the direction of this field.

Uhlenbeck and Goudsmit applied to this rotation of the electron about itself the same quantum laws which Bohr had introduced in treating the revolution of the electron in its orbit about the nucleus. They deduced that all electrons always have the same fixed spin velocity. If one chooses as the unit of measurement the value h divided by 2π (where h is Planck's constant and π is the well-known number 3·14) one finds that the electron always has a spin equal to $\frac{1}{2}$.

Furthermore the application of quantum laws to the rotation of the electron about itself leads to another very important conclusion. I have said that as the electrons are electrically charged and are rotating about themselves they behave like small magnets; and thus if they are put in a magnetic field their axes tends to orientate themselves with respect to the direction of the field. By applying quantum mechanics Uhlenbeck and Goldsmit deduced that the spinning electron could adopt only two positions with respect to the field. In both cases the axes of spin would be aligned with the field's direction but in one case the spin of the electron would be clockwise, in the other case anticlockwise. That is, the spin could have only two values, $+\frac{1}{2}$ and $-\frac{1}{2}$. Thus if we put many electrons in a magnetic field—even a weak magnetic field—they would all become orientated with their axes parallel to the field but for some the axis would be in the same direction as the magnetic field and for others in the opposite direction. The values of the spin and of the magnetic moment of the electron deduced in this way by Uhlenbeck and Goudsmit permitted a quantitative interpretation of the optical and magnetic properties which had not been explained in Bohr's atomic model.

But perhaps the most interesting development occurred some years later when Dirac introduced his theory which combined relativity with quantum theory; he discovered that the electron must essentially possess a spin and a magnetic moment having just the values that had been deduced by Uhlenbeck and Goudsmit starting from their hypothesis of the rotating electron.

This hypothesis has been a very important step in the history of our understanding of the atomic world. It has provided a model with which we can attribute to the electron a spin and a magnetic moment. We now know that these quantities and their numerical values are essential properties of every electron without however being obliged to think that the electron really revolves about itself. One sometimes, however, refers to this now superseded picture of the spinning electron in order to describe intuitively, even though in an incorrect manner, the properties of the electron.

We will see later that the property of having a spin is not a characteristic only of electrons; one can define for any particle a spin whose value is one of the essential characteristics of the particle.

PAULI'S EXCLUSION PRINCIPLE

Since in the atom of hydrogen there is only one satellite electron this electron must naturally revolve alone in the orbit corresponding to its energy. In normal conditions it will revolve as we know in the orbit closest to the nucleus—the innermost orbit.

But when we go on to consider atoms of elements heavier than hydrogen, that is of elements with atomic numbers greater than one, there will be more than one electron revolving round the nucleus and, as we now well know, the number of these electrons is equal to the atomic number of the element (a number which is equal to the value of the positive electric charge of the nucleus if one takes as the unit of charge the value of the charge of an electron). For example in helium there are two electron satellites. In what orbit will these two electrons rotate in normal conditions, that is when the atom is in its ground state? One might think that both electrons would rotate in the innermost orbit, and in the same way one might think that the same applied to the three electrons in the atom of helium which is in the ground state, and so on for all the other elements.

But this is not so. In 1925 the German physicist Pauli discovered, by studying the spectrum of alkaline metals, that *in each orbit there could revolve only one electron*.

This is found to be true in all atoms so it has been taken as a fundamental principle of atomic physics and has been given the name of *Pauli's Principle* or *Exclusion Principle*. It has been very fruitful in results and has supplied the key to the interpretation of that fundamental natural law which is Mendeleev's periodic system.

We recall that the orbits in which the electrons can travel are grouped in layers (K, L, M, ...) and that the innermost layer K is made of two orbits, the layer L of eight, the layer M of eighteen orbits, etc.; but by Pauli's Exclusion Principle only one electron can revolve in one orbit. Thus the layer K can contain only two electrons, the layer L only eight, the layer M only eighteen, etc. We recall also that in normal conditions, that is when an atom is in its ground state, the planetary electrons try to occupy the innermost orbits, that is those of lowest energy. But by Pauli's Principle they can only do this when these orbits are not already occupied.

Thus in normal conditions in a hydrogen atom the single electron moves in one of the two orbits of the K layer. In following elements of atomic number 2, that is in helium, the two electrons go to occupy the only two places which exist in the layer K which is therefore full. Helium then is an element whose atom has a completed electron layer. This arrangement

Atomic Physics

is particularly stable. It means that it will be difficult for helium to take part in a chemical reaction. Similar behaviour will be shown by all those elements whose atoms consist of a closed ensemble of a nucleus with one or more complete electron layers—for example neon with full K and L layers (ten electrons equals two plus eight), etc. The elements helium, neon . . . have similar chemical properties. They are the noble gases.

In lithium which has the atomic number 3 the third electron, not being able to find a place in the first layer, K, which is already full, will rotate on the innermost orbit of the second layer, L. This third electron however will be less strongly bound than the two others since it moves in a wider orbit. The lithium atom therefore has a certain tendency to lose this electron and so become changed into a monovalent positive ion. Similar behaviour will be found in sodium which has eleven outer electrons of which the outermost revolves around a stable ensemble consisting of the nucleus and of two complete layers; potassium, rubidium, etc. which all have a less strongly bound outermost electron, all behave in the same way. These elements, which have similar chemical properties, are the alkalis.

And so it goes on. . . . Let us consider now fluorine, the element with atomic number 9. Of these nine electrons two make up the K layer and the remaining seven move in the L layer. Now as this layer could contain eight electrons and as every full layer has a particular stable structure this atom will have the tendency to acquire another foreign electron in order to make up the layer; it will then become a monovalent negative ion. Chlorine, bromine, iodine, etc., behave in a similar way; they are all elements in which the atom has a complete external layer but for one missing electron. These elements are the halogens.

Let us now see what happens, for example, when an atom of sodium joins with an atom of chlorine in order to make a molecule of sodium chloride. As we have seen, the sodium atom (whose symbol is Na) has the tendency to lose its external electron thus becoming a positive ion Na^+; and chlorine has the tendency to acquire an electron and thus become a negative ion Cl^-. These two ions, having electric charges of the opposite sign attract each other and join together to form a molecule of sodium chloride which is electrically neutral.

It is not in all compounds however that the binding between the atoms is of this type (ionic binding). In other cases it is of a much more complicated nature.

The electron orbits become more dense as one moves from the inside to the outside of the atom; that is, the difference in energy existing between the inner orbits is greater than the difference in energy between the outer orbits. Thus a quantum emitted from the atom when an electron moves from one to another of the innermost orbits has a greater energy than a quantum emitted when an electron moves between two external orbits. Thus by Planck's relation an electron which jumps between two outer orbits emits an electromagnetic wave of lower frequency than the frequency of the wave emitted when an electron moves from one inner orbit to another.

The atom can in this way emit a huge range of electromagnetic radiation: from light rays (emitted when an electron moves between outer orbits) up to X-rays emitted when an electron moves between inner orbits passing via ultra-violet radiation.

If an external disturbance supplies to the atom a relatively small amount of energy enough only to make one or two of the electrons which are farthest from the nucleus jump to orbits which are still farther out, then these electrons, on returning to their stable position will emit energy in the form of light. If, on the other hand, the disturbance supplies sufficient energy to extract from the atom one of the innermost electrons another electron will immediately jump into the vacant position and an X-ray will be emitted.

As we have already noted, in general an external agent which excites one atom in a body, that is which causes one electron in an atom to move in an orbit outside its normal orbit excites also the other atoms making up the body; so that when these electrons return from their unstable position to their normal orbit they emit a very large number of quanta (as many quanta as there were excited atoms). Now all these quanta are equal since they all have the same energy. Thus the radiation which is emitted has a well-determined frequency. For example a ray of red light is made up of identical quanta each having the same energy which corresponds to the frequency 4.3×10^{-14} vibrations a second.

This is the way in which all forms of electromagnetic radiation are emitted, heat rays (infra-red radiation), light and X-rays. The emission of gamma rays, which are electromagnetic radiations of a frequency higher than that of X-rays occurs, as we shall see later, by a mechanism which although different is analogous to that which we have just described.

With the Bohr–Sommerfeld model one is able to make a logical theory of the distribution of the electrons in the various layers which accounts very well for the chemical and spectroscopic properties of the various elements. The periodic table discovered by Mendeleev in 1868 on a purely empirical basis can now be neatly interpreted and justified on the basis of this simple model.

Despite this however it was soon realized that the Bohr–Sommerfeld theory, even with the introduction in 1925 of the hypothesis of the spinning electron, could not interpret all the experimental results. Theoretical physics would very soon require a new orientation and its fundamental principles will undergo profound revision.

But before looking at the new orientation assumed by theoretical physics after 1925 I will stop to explain the importance of the Bohr–Sommerfeld theory in the story of the study of the atom.

VALUE OF THE BOHR-SOMMERFELD THEORY

The atomic theory of Bohr–Sommerfeld has an enormous importance in the history of atomic physics. It dominated atomic physics until 1925 guiding an enormous amount of work both theoretical and experimental; with it, as we have seen, it was possible to interpret and to predict many of the phenomena of atomic physics.

However, there are many serious objections to be made to the Bohr theory. First of all, it failed to satisfy what should be fundamental conditions for any theory: that laws, once they have been formulated, should always be found to be valid and should be adequate on their own to interpret phenomena. During the life span of Bohr's theory not only were the original laws of quantization modified in order to account for experimental results but new laws and hypotheses were introduced which had not been suggested by the original postulates.

Furthermore, Bohr's theory was only able to interpret satisfactorily the spectrum of the hydrogen atom and of other similar atoms. It was not even able to explain the spectrum of the atom of helium—the simplest pattern after that of hydrogen; and finally Bohr's theory did not supply any means for predicting the intensity and the polarization of spectral lines. The only thing which it could do in this regard was to go back using the correspondence principle to classical mechanics and electromagnetism.

Even leaving aside these difficulties and failures Bohr's theory was open to serious criticism. It based itself, as has been said, on 'a strange alliance between the concepts and the formulae of classical dynamics and the quantum method'. The introduction of discontinuity seemed so arbitrary that the Bohr–Sommerfeld theory 'has never been' in the words of Persico 'regarded as the definitive expression of a physical theory but rather as a provisional codification in a manner of speaking of the ruptures to be made in classical mechanics and electromagnetism in order to make them fit the atomic world'.

Bohr himself was perfectly aware of all this. From the beginning he continuously underlined the provisional character of his atomic model. It was necessary to open new ways which were completely independent of classical theory.

These new ways will be opened and followed and new theories will be developed; but although the new theory will be based on concepts very different from those of Bohr's theory it will yet be from many points of view a refinement of it. Bohr's theory of the atom can be considered as an intermediate theory which links the old ideas to the new. Here once again science, as on many occasions in its history, succeeds by successive approximation.

NEW QUANTUM THEORIES

When physicists realized that the Quantum Theory of Bohr and Sommerfeld could only be a first step towards a complete understanding of atomic phenomena they began to try to construct a more refined theory. About 1925 most physicists became convinced that this could be done only by a fundamental revision of the principles of theoretical physics. They realized that, once again quoting Persico, 'this critical situation arose from the assumption inherent in all the previous theories that atoms, protons and other entities of the atomic world should be conceived as particles or exceedingly small mechanisms although subject to laws which were different to those of ordinary bodies. They might, on the other hand, be entities which are essentially different to which one cannot apply (and indeed, there is no reason why one should apply) our intuitive concepts of body, movement, etc. In other words, the possibility was considered that the atomic world although subject to laws which could be mathematically expressed could not be represented by any intuitive model.'

After 1925 theoretical physics adopted a new approach which is given the generic term of Quantum Mechanics. It developed in three different ways. *The Wave Mechanics* of de Broglie, the *Matrix Method* of Heisenberg and the *Operator Method* of Dirac and Jordan.

We must bear in mind that whilst as we have already said, this new general outlook of theoretical physics (in its three forms) is called Quantum Mechanics, the Bohr-Sommerfeld theory is often called the Classical Theory of Quanta. Both then form part of the Quantum Theory which includes all the theories in which Planck's Constant h plays an essential part.

The three different forms of quantum mechanics are not mutually contradictory. It has been said that they differ only 'in the conceptual ornaments with which they are adorned'; they are mathematically equivalent; one may prefer to use one rather than another. They complete the classical theory of quanta just as it completes classical physics. For large enough quantum values the new Quantum Mechanics merges with the classical theory of quanta; for still larger quantum values it merges with classical mechanics. As we have already seen science moves forward by successive approximation.

I will dwell briefly on the first two forms of Quantum Mechanics—briefly because their complex mathematic symbolism does not allow us here to go into greater detail. I will not discuss Dirac's Operator Theory because as it is mathematically extremely abstract it cannot be explained by familiar concepts.

I will consider first Wave Mechanics.

LIGHT

Already in 1905 following on the success of Planck's hypothesis physicists began to realize that the laws of classical mechanics could not be applied to the inside of the atom; and in that year Albert Einstein considering Planck's laws made a theoretical study of the nature of light (and in general of the nature of radiation). He reached a conclusion which appeared paradoxical.

What is the nature of light? 'What is this thing which we call light?' writes Ronchi, 'which travels in straight lines, is reflected, is refracted, which draws the colours out of bodies, which carries with it heat, which forms images through lenses, and which is recorded on the retina of the eye.' This problem was tackled by the physicists of the seventeenth century, including the two greatest, Newton and Huyghens.

In 1690 the Dutch physicist Huyghens, who was the inventor of the pendulum clock, published at the age of more than sixty his most outstanding work. It was a thesis on light in which it was, for the first time, stated clearly that light is a vibration similar to sound; with this hypothesis he found the laws of reflection and refraction and interpreted the phenomenon of double refraction. But this idea of the wave nature of light appeared so strange that it was bitterly disputed. Newton himself did not accept it and put forward his own corpuscular theory. According to this theory light is made up of very fast and very small particles which travel in straight lines in any homogeneous and transparent material. Newton did, however, attribute to these particles certain wave properties.

For two and a half centuries there was a conflict between these two conceptions of light and the fact that, as we shall see, neither of the two has finally triumphed over the other constitutes one of the most curious results in the history of physics.

It was the corpuscular theory which, for more than a century, had the greatest influence not only in the scientific field but also amongst the general public. It was only in 1815 that the French physicist Fresnel became convinced that certain phenomena (the phenomenon of interference observed by Young fourteen years beforehand and that of diffraction which had been observed by

Atomic Physics

Father Grimaldi in the seventeenth century) could only be explained according to the wave theory of light which had been put forward by Huyghens. Even if it was necessary later to abandon the simple pictures (such as that of the ether) on which Fresnel based his inspired theory the mathematical formulation of the wave theory is still completely valid.

Thus in the first half of the nineteenth century Newton's corpuscular theory yielded first place to the wave theory of Fresnel which was able to explain correctly all the phenomena then known; it explains in particular one apparently paradoxical fact which appears in interference phenomena; that is, that in certain special circumstances light added to light can give darkness.

If we allow to fall on a screen two narrow rays of light both of which have the same frequency but which have travelled through different distances we see on the screen not one completely illuminated zone but a series of zones alternately light and dark. Now it is impossible to explain this fact starting from the hypothesis that light is made up of particles; for with this hypothesis 'adding light to light' could only lead to an increase in the illumination just as firing simultaneously with two machine guns on the same target (using a well known example) would result in the addition of the effects of the individual machine guns. But if light is a wave phenomenon what happened on the screen would be analogous to that which one sees on the surface of a pond into which are thrown simultaneously two stones which are separated by a certain distance. From the two points on the surface of the water struck by the stones there would emerge a series of small waves. In those places where the crest of a wave from the first stone met a crest from the second stone the two heights would be added; but where a crest from the first wave meets a trough from the second the two effects cancel and there is no displacement of the water.

This is a very schematic explanation of the phenomenon of the interference of light which had been discovered by Young in 1800. It is associated with an assembly of phenomena (which are called diffraction phenomena) which take place when a ray of light encounters on its path a screen or a slit or a hole etc. whose dimensions are so small that they are of the same order of size as the wave length of the light, i.e. about one ten-millionth of a millimetre. These phenomena were very well explained by the wave theory of Fresnel.

However, the details of the way in which light is produced were not, at that time, known. It was only in 1865 that the British physicist Maxwell—who was able to sum up in a single system of equations all of the laws of electricity—deduced that electromagnetic waves must exist. He wrote: 'If it shall be found that the velocity of propagation of electromagnetic disturbances is the same as the velocity of light ... we shall have strong reasons for believing that light is an electromagnetic phenomenon ...' All subsequent measurements have completely confirmed the electromagnetic theory of light put forward by Maxwell.

Twenty years later a German, Hertz, demonstrated by experiment the electromagnetic nature of light thereby confirming Maxwell's interpretation; the door was then opened to a long series of experiments and of inventions which led to the tremendous growth of radio communication.

Wave optics has been tested in an enormous number of experiments; until the first years of the present century all of them confirmed the wave character of light. Nobody would have imagined that very soon the validity of this wave theory would once again be disputed. This, however, happened when Einstein who was studying the laws of a curious phenomenon (the photo-electric effect) was obliged to put forward, even though in a new form, a corpuscular theory of light.

THE PHOTO-ELECTRIC EFFECT

I have said that in 1905 (five years after Planck had put forward his hypothesis that energy is emitted and absorbed not continuously but in quanta) Albert Einstein was obliged to accept that not only was energy emitted and absorbed in quanta but that also it was localized in granules during its journey through space. To these granules of energy he gave the name of light quanta or photons.

Einstein reached this conclusion as the results of a series of theoretical arguments based on Planck's laws. In support of these theoretical arguments he showed how the so-called photo-electric effect which had been first observed by Hertz in

1887 could be logically interpreted by his hypothesis that light travels as photons whilst it was impossible to explain it starting from the wave theory of light.

The photo-electric effect essentially consists in this: if one allows a ray of light of very short wave-length (ultra-violet rays, X-rays or gamma rays) to fall on a metal surface the surface will emit electrons. The electrons are expelled with a certain energy which they have evidently received from the incident radiation. Now one finds that the energy—that is, in the final analysis, the velocity—with which these electrons are expelled from the metal is independent of the intensity of the incident radiation but is increased if the frequency of the radiation is increased. This means that the electrons which are expelled when the metal surface is struck by gamma rays have a higher velocity than that of electrons emitted as the result of illumination by X-rays; and in their turn electrons from X-rays have a higher velocity than those produced by illumination by ultra-violet radiation; this is because the ultra-violet rays, X-rays and gamma rays are electro-magnetic radiations of increasing frequency. But it is also found that the velocity of the electrons which are expelled when a metal surface is struck by radiation of a fixed frequency is independent of the intensity of the radiation.

Now this independence of the velocity of the electrons on the intensity of the incident radiation cannot be explained in terms of the wave theory of light; for according to this theory the energy of an incident wave increases with its intensity. Thus radiation of the same frequency—for example ultra-violet light of the same frequency—which is more intense should give rise to electrons of higher velocity; this is just the opposite of what is observed. Another fact which conflicts with the wave theory of light is this. One can direct on to a metal radiation which has such low intensity that its energy is very much less than the energy necessary to tear electrons from the atoms of the metal; but even in this case electrons are expelled with the same velocity which corresponds to the frequency of the incident radiation.

In a celebrated article published in 1905 Einstein showed that the behaviour of the electrons in the photo-electric effect, which appeared incomprehensible in terms of the wave theory of light, could easily be explained if a corpuscular theory of light was adopted and light was assumed to be made up of photons.

Every photon of frequency ν has an energy given by $h\nu$ where h is Planck's constant. When one of these photons strikes the electron of an atom in the metal the energy is given to the electron which in this way is freed from the atom to which it belonged and escapes from the metal surface. The statement that a quantity of light of a certain frequency is more intense than another quantity of the same frequency means only that it is made up of a larger number of photons; but in the two cases each individual photon has the same energy. This is why the velocity of the electrons which are torn from the metal is independent of the intensity of the incident radiation provided that the frequency is fixed. One sees that what varies in this case is only the number of the electrons which is increased according to the intensity of the incident light; this can be well understood since as the intensity is increased so is the number of incident photons.

In order that the velocity of the electrons should be greater it is necessary that the quanta should have a higher energy which means by Planck's relation that the frequency of the incident radiation should be higher. When this frequency is less than a certain value the energy of each photon will be too small to extract an electron from the metal; and for this reason the photo-electric effect does not occur if the metallic surface is illuminated by normal visible light whose frequency is too low.

Thus Einstein's theory of photons accounts perfectly for the phenomena observed in the photo-electric effect and these phenomena cannot be explained in terms of the wave theory of light. The existence of photons received subsequently further confirmation from other experimental facts.

THE WAVE THEORY AND THE CORPUSCULAR THEORY OF LIGHT

On the one hand interference phenomena could be explained by the wave theory of light and could not be explained by the hypothesis that light was of a corpuscular nature; on the other hand, other phenomena and principally the photo-electric effect showed that light had a discontinuous structure of quantum origin, i.e. that it was composed of photons. Thus in the theory of light there were two models, one wave and one corpuscular, each of which could explain one set of phenomena but were incompatible with the other set. How could this conflict be eliminated?

Before showing how this serious situation was resolved it is necessary to underline the fact that Einstein's theory of light quanta is not a purely corpuscular theory of radiation in the sense that it claims that light is made up of material particles as Newton had believed; for in the words of de Broglie, 'the very way in which Einstein defined the light quantum introduces a non-corpuscular element, the frequency. A purely corpuscular theory of radiation could not allow the definition of something periodic, like frequency and in fact the frequency which appears in Einstein's definition is that of the wave theory, that which is deduced from interference phenomena.... The relation which defines the energy of the photons as equal to the product of the frequency by Planck's constant cannot therefore provide the base for a purely corpuscular theory of radiation. It establishes rather a type of bridge between the wave aspect of light ... and the corpuscular aspect which was revealed by the discovery of a photo-electric effect.' And further on, 'the very form of Einstein's relation invites us to associate the corpuscular and wave idea in such a way that both the terms in the relation have physical significance'.

Thus physicists accepted the duplicate nature of radiation and according to the phenomena which they were studying used either the wave theory or the corpuscular theory. It was a very unsatisfactory situation; in the words of W. Bragg, 'On Mondays, Wednesdays and Fridays we use one hypothesis; on Tuesdays, Thursdays and Saturdays we use the other.'

WAVE MECHANICS

De Broglie started from a consideration of this 'terrible dilemma' which was revealed by the investigation of the nature of light (wave or particle?); between 1923 and 1924 he arrived at the first formulation of his theory.

De Broglie was struck by the fact that, in his own words, 'the relationship between frequency and energy which Einstein placed at the basis of his theory of photons intimately linked the dualism (wave particle) of radiation to the existence of quanta'. He asked himself whether this dualism was not an expression of the nature of Planck's quanta; whether, in other words, 'one should not find the same dualism wherever Planck's constant is present'.

Now according to Bohr's theory the quantum of Planck appears in the properties of the electron (as is shown by the fact that an atom can exist only in certain fixed energy levels); and so de Broglie asked himself, 'does the electron not show a dualism analogous to that of light?' and in general should one not attribute a wave aspect to material particles. He thought that even particles and, in particular, electrons should possess some wave-like characteristics.

This idea appeared terrifyingly bold and almost fantastic. The electron had never displayed wave properties and had always in macroscopic phenomena behaved as a material point; but the fact that in atomic phenomena it does not follow the laws of classical mechanics shows that its properties could not always be those of a simple particle and so this idea which seemed bold and fantastic appeared to de Broglie as 'urgent and fruitful'. The subsequent development of the idea has revealed the accuracy of de Broglie's first intuition.

De Broglie proposed then to attribute to electrons and in general to all of the so-called particles (that is to the constituents of atoms, electrons, protons, alpha-particles) 'a dualistic nature analogous to that of the photon giving to them a double aspect, wave-like and corpuscular based on Planck's quantum'.

Thus, guided by the association of photons with waves of radiation de Broglie made the hypothesis that there should be an analogous association between material particles and a certain periodic phenomenon, a new type of wave (a de Broglie wave). The nature of these waves remained, however, extremely mysterious.

For a particle of mass m moving in a given direction with a velocity v the de Broglie wave had a wave-length equal to h divided by mv where h is Planck's constant. This showed that in an imaginary world in which Planck's constant was very much smaller the same particle would be associated with a

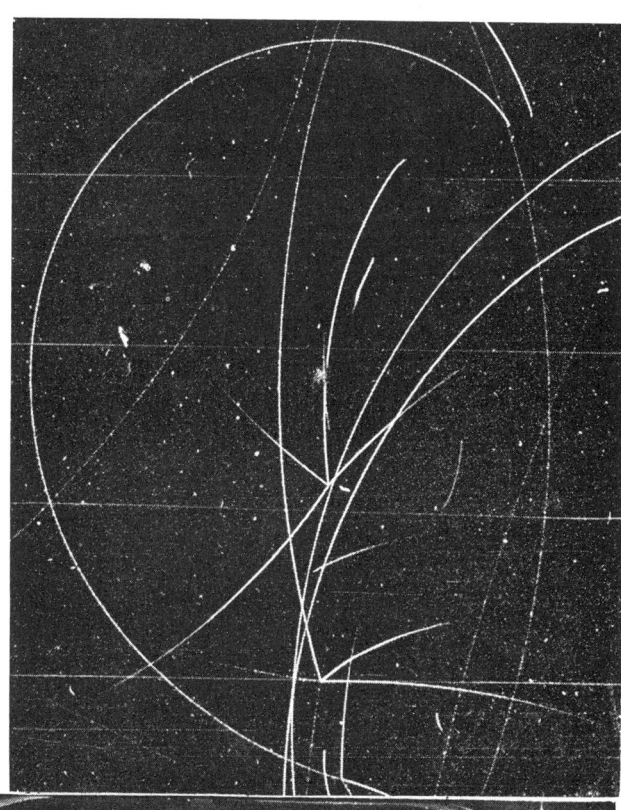

3a Tracks of protons and electrons in a Wilson chamber with magnetic field

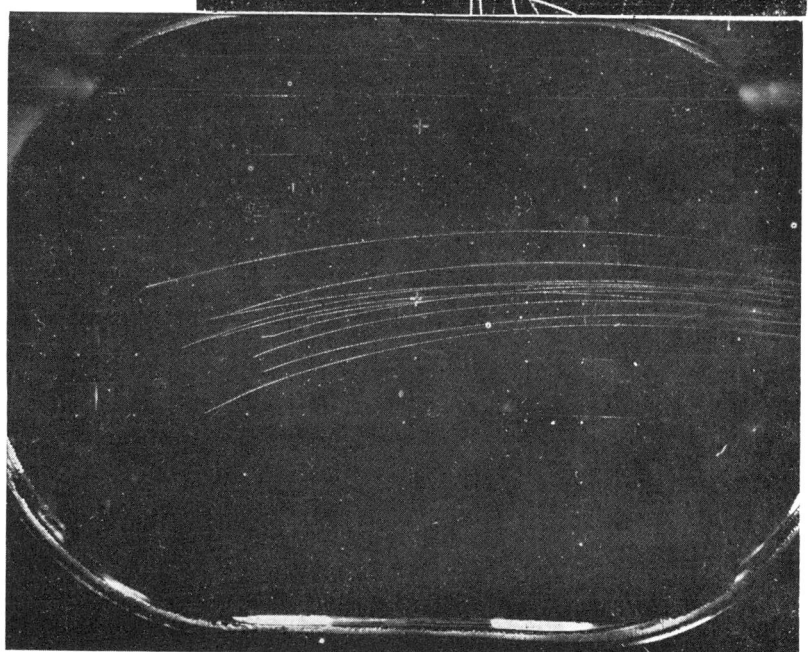

3b Tracks of protons in a bubble chamber

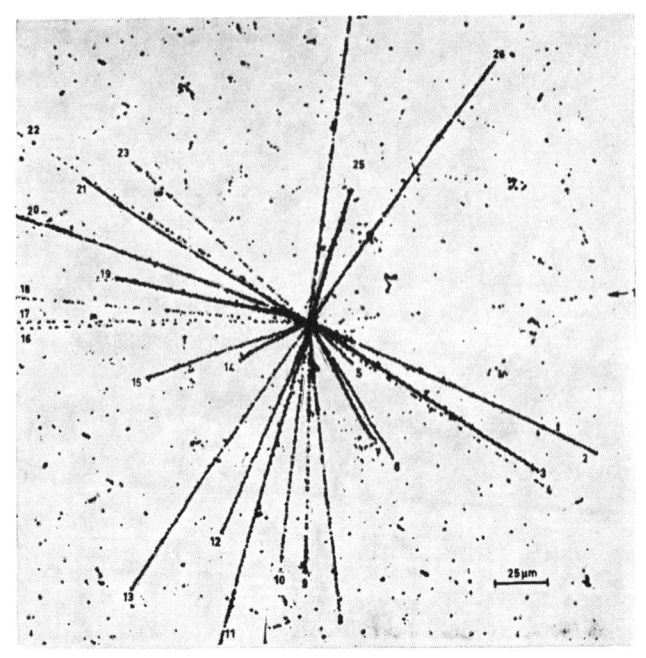

4a Microphotograph of the disintegration of an atomic nucleus produced in a special photographic emulsion by a high energy particle (track 1 on the right-hand side of the photograph)

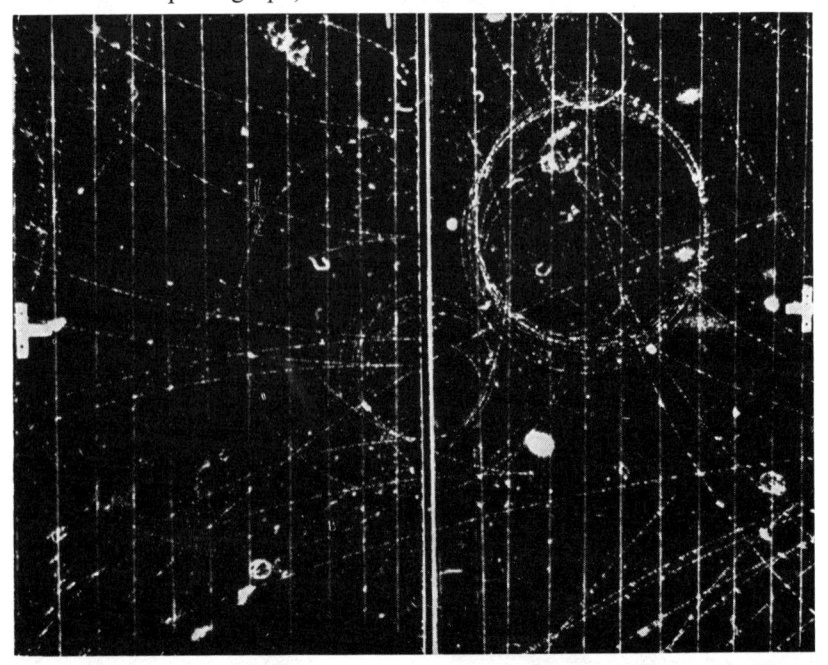

4b Pairs of electrons and positrons produced by gamma rays of 335 Mev

wave of very much smaller wave-length; that is, the wave properties of matter would be even less evident than they are in our world where h has the value of $6 \cdot 6 \times 10^{-27}$ erg seconds. Thus the wave properties of matter are fundamentally only an expression of quanta; in a world in which h was equal to zero they would not exist.

De Broglie not only suggested these ideas but brought them together in a theory which gave a mathematical formulation of the new waves which he postulated. He showed how they should behave and what phenomena they should produce.

But de Broglie's theory in this first form led to the same quantization condition as that of Bohr and thus like Bohr's theory was not always able to account for the observed positions of the spectral lines. In 1926 the German physicist, Erwin Schrödinger, developed de Broglie's theory and examined it in a new light. He was the first to deal rigorously with the problem of the movement of the de Broglie waves, and he wrote down explicitly the equations for the waves of wave mechanics. He calculated the position of the spectral lines and attained results which in many cases were different from those attained from the classical theory of quanta and were in better agreement with the experimental results.

A critical check of de Broglie's theory was made in the experiments carried out in 1927 by Davisson and Germer. These physicists discovered that if a beam of slow electrons is directed on to a solid crystal (that is a solid in which the atoms are arranged in a regular pattern) diffraction phenomena were produced analogous to those which occurred when the crystal was struck by a beam of X-rays, that is of electromagnetic radiation whose frequency is higher than the frequency of light. This was a spectacular demonstration of the validity of de Broglie's hypothesis which had attributed a wave aspect to electrons; but this was not all. Calculations made on the basis of the experimental results of Davisson and Germer gave for the wave length associated with the electrons just the value which had been found theoretically by de Broglie. Other physicists immediately carried out similar experiments and found that the predictions of de Broglie were always confirmed.

I have said that the particles which are associated with waves are the extremely small particles which make up atoms: electrons, protons, alpha-particles. Now why is this?

In theory a de Broglie wave is associated with every material particle which moves; thus, one should observe diffraction effects for all types of particles, including for example grains of sand; but according to de Broglie's theory for a given particle velocity the associated wave-length is smaller the larger the mass of the particle. Thus only for very minute particles with extremely small mass (and these are just atomic particles) is the associated wave-length large enough to be able to give rise to the phenomena which we can observe with our macroscopic apparatus.

All electromagnetic radiation of whatever wave-length and every particle of whatever mass, present this double aspect, wave and particle; but only for electromagnetic radiations of very small wave-lengths (ultra-violet rays, X-rays, gamma rays) and for particles of extremely small mass (protons, electrons . . .) is our experimental equipment adequate to supply evidence both of the corpuscular aspect and of the wave aspect.

The wave properties of light were discovered some centuries ago and the corpuscular properties were only discovered much more recently. For electrons (and in general for material particles) it was the other way round and this is the reason we are accustomed to think of radiations and of light in particular from the wave aspect and the particles from the corpuscular aspect. And so we used to believe that in nature there existed radiation made up of waves and radiation made up of particles but we now see that all radiation exhibits according to the circumstances both wave properties and corpuscular properties.

We shall see later how the new wave mechanics resolved this wave-particle dilemma which is exhibited both by light and by matter.

DE BROGLIE WAVES

As d'Abro has shown very clearly, as long as the wave-particle dualism was limited to radiation one might think that it displayed a new property peculiar to the strange thing which we call light, but when after de Broglie's contribution the same

Atomic Physics

dualism was extended to matter physicists realized that they were confronted with some general principle of hidden significance. Various attempts were then made to explain this dualism.

The first was made by de Broglie himself. He supposed that only the waves were real and that the corpuscular aspect was due merely to some characteristic of the waves; that is, that the dual aspect of light and matter was not due to the fact that waves and particles exist independently in nature; but the waves were the fundamental quantities and the particles were, in a manner of speaking, only by-products.

This interpretation met with various difficulties and had to be rejected. De Broglie then tried again and put forward a new interpretation which was called the hypothesis of pilot waves: Both particles and waves have a physically independent existence. The wave does not determine the trajectory actually followed by a particle but it 'guides' the particle along one of the trajectories which are associated with every wave phenomenon. The probability that the particle should be at a given time at a given point is proportional to the intensity which the wave had at that point and at that time.

But not even this explanation, although it was very appealing, was able to withstand criticism; amongst other things, it was in certain conditions in contradiction with the theory of relativity; but although it had, therefore, to be abandoned it is, nevertheless, of particular interest. It made clear the great difficulties which one encounters when one wants to allot to a particle a well-defined trajectory and a well-defined momentum. It led to the birth of a suspicion that a particle with a rigorously defined position and with a rigorously defined momentum might be 'a myth'.

Various other attempts were made to explain the relationship between particles and waves. The last of these which was initiated by Born provided the basis for the uncertainty relationships of Heisenberg to which I shall refer later. It is not possible to reproduce here Born's arguments. He started from two fundamental hypotheses; the first concerns the probability that a particle will be at a given point at a given time; the second refers to the probable value of the velocity and of the energy of the particle. The waves do not represent physical quantities;

they are only symbolic and can be regarded as probability waves. They define the probability of future events which can be derived from our knowledge (which may be precise or which may be vague) of the present conditions. Although the waves associated with the particles have no physical reality they must be taken into consideration because only via the wave representation can one succeed in correctly predicting possible movements of the particles.

And so in this interpretation the waves must be considered as symbolic whilst the particles represent the real world. But if one analyses rather more accurately the phenomenon of interference, and if in particular one considers the case of radiation from a source of light so weak that one can consider the arrival of individual photons on a screen, one encounters a paradox as long as one continues to consider the particles (photons, electrons, etc.) as minute bodies in the sense of classical mechanics. In other words, the paradox arises if these particles have at a given time a fixed position and if they move along a fixed trajectory which could be calculated from the knowledge of the original position of the particle, its mass, its velocity and all the other experimental conditions. The paradox was eliminated only when the physicists Heisenberg and Bohr, abandoning intuitive ideas, made the hypothesis that photons, electrons, etc. are not particles in the ordinary sense of the word. They are *not* localized at a point in space at a given time; for none of them can one speak of a trajectory defined as the assembly of all the points successively occupied by a particle in the course of its movement.

We shall discuss immediately the sources, the development and the results of this theory of Heisenberg and Bohr which is today accepted as valid, but I should first make one point clear. According to the last interpretation of the wave-particle dualism (that of Born, to which I have referred above) the waves are only waves of probability and should be regarded as symbols whilst the particles correspond to the real world. But in the theory of Heisenberg and Bohr the waves and particles are put on the same level, both are, in a certain sense, symbolic, in the sense that they do not exist independently but are two different aspects of the same reality. According to the experiments which we make we encounter one or other of these aspects. A body, for example a photon, can, in a manner of speaking, spread itself over an extended region of space (and thus give rise to interference phenomena) and in this way it shows its wave aspect; but if we want to observe its condition at a given time (for example, when collected on a screen) this spreading of the particle stops immediately; it condenses, displaying its particle aspect.

We shall now go into rather more detail; and we shall see how these ideas are justified in their logical derivation and are substantiated by experimental results. They can appear very strange to our brains which are accustomed to the concepts of classical physics; but we must always remember that we cannot expect to be able to continue to use familiar notions and concepts when we move into the world of microscopic physics.

There is no doubt that in the wave-particle dualism physicists encountered one of the most difficult problems which they had ever met.

HEISENBERG MECHANICS

In 1925 de Broglie created the concept of a wave associated with a material particle thus, 'suddenly opening a royal road to the theory of quanta' (Manneback); and the German physicist Heisenberg created a new system of mechanics which had the very great merit, compared to the theory of Bohr-Sommerfeld, of being a perfectly coherent theory from the logical point of view; he started from an organic system of postulates and deduced by one consistent method all the results. The technique used by Heisenberg is the so-called *theory of matrices*, a part of mathematics which, though it had long been known to mathematicians, had not yet been applied to physics.

The use of this complex mathematical technique makes it impossible for those who have not a deep knowledge of mathematics to understand fully the quantum mechanics of Heisenberg. It is, however, possible to explain the ideas from which Heisenberg started in order to build his theory and to say something about the far-reaching revision and logical clarification of the bases of physics to which it led and finally to discuss its philosophical significance. Heisenberg placed at the basis of his theory a fundamental, logical principle which had already over the course of several years been gradually established; we have already seen a particular case of this principle when I referred in the previous paragraph to the trajectory of a particle. The principle states that one should not introduce into physics magnitudes which cannot be measured by means of an experiment, an experiment which can be executed—or if it cannot be executed because of practical limitations of the apparatus available is at least in principle feasible. On the other hand we should attach no physical significance to a quantity which could only be measured by an ideal experiment, an experiment which is infeasible, not because of practical limitations of the available apparatus, but because it is logically inconsistent or conflicts with a law of physics.

An ideal experiment may be impossible in practice or theoretically impossible. I will give an example to make clear the difference between these two types of ideal experiment. One might think of exploiting the enormous amount of heat contained in the sea by taking from it a minute part of its heat so that its temperature would drop by a fraction of a degree; one might then think of using this heat to operate for example a steam engine and thus obtain mechanical work. Now this ideal experiment cannot be made not because a suitable apparatus has not yet been invented or because for practical reasons we have not sufficient means to construct it. The experiment cannot be made because it is in principle impossible. It violates the second law of thermodynamics which states that there is no process which has as a single final result the transfer of heat from a cold body to one which is warmer. One might, on the other hand, construct an apparatus which could use heat from the sea to warm a body which is at a lower temperature than the sea, for example, to dissolve a quantity of ice. The construction of such an apparatus might not be convenient from the economic point of view and might be complicated by practical difficulties, but it is an experiment which is in principle possible because it is not forbidden by any law of physics.

Thus every concept which is employed in physics should be defined by what is called an operational definition, that is by specifying a series of physical operations which are in principle possible. One must carefully analyse every argument used in theoretical discussion in order to make certain that it does not

Atomic Physics

employ concepts which cannot be defined by operational definitions.

This principle, as one can easily understand, profoundly affects the method of argument itself and it goes to the heart of the problem of knowledge. It has had in the recent history of physics very impressive results. In fact, it suggested to Einstein the observation which was the starting point of his theory of relativity. Einstein asked himself, 'What does it mean to say that two distant events occur simultaneously. What, for example, does it mean to claim that on the earth and on a very distant star two events occur at the same time.' Einstein pointed out that this concept of the simultaneous occurrence of two events which were separated in space is significant only when one can describe the method which allows one to decide whether the events are in fact simultaneous. It was from this observation that Einstein developed into a logical and coherent system his theory of relativity.

In 1905 Einstein published his first work on the theory of relativity. Twenty years later Heisenberg applied the same principle to the field of atomic physics and thus started that critical revision of concepts which has profoundly modified our ideas on the constitution of matter.

Heisenberg then set himself to formulate the laws which control phenomena in the inside of an atom in such a way that only quantities which could be measured entered into the calculations. The phenomena which take place in the inside of an atom are displayed to us by means of the spectrum which the atom emits; what we can measure are the positions and the intensity of the lines of the spectrum of that atom. Thus Heisenberg formulated the laws which regulate such atomic phenomena by linking them directly only with the intensity and the frequency of the radiations emitted by atoms; and in order to link together these observable quantities he made use of a complex mathematical technique to which I have referred and which is called matrix algebra.

This was the original of the matrix mechanics of Heisenberg. Subsequent work by Heisenberg himself, by Born, by Jordan and by Dirac 'have quickly produced a system', according to Bohr, 'which for coherence and generality can compete with classical mechanics'.

After Heisenberg had published the results of his matrix mechanics other physicists began to travel along the road which he had opened. In the meantime Schrödinger (who, taking up again the idea of de Broglie, had explicitly written the wave, equation of quantum mechanics and who had succeeded in predicting various experimental results) realized that the two mechanics which started from such different points of view—wave mechanics and the matrix methods of Heisenberg—led in every case to the same results; in a celebrated article he showed that they are mathematically different forms of the same theory. Each problem can be treated by one method or the other according to the convenience and the mental outlook of the research worker. The matrix method is the faster whilst the method of wave mechanics is more easily handled.

THE UNCERTAINTY PRINCIPLE

The application to the field of atomic physics of the principle that every concept which is introduced must be defined by an operational definition led Heisenberg to present the Uncertainty Principle. This principle has had an enormous influence on our ideas and has allowed us to understand the deep significance of the new mechanics.

The rigorous application of the concept of operational definitions obliges us to give up in the study of atomic phenomena certain ideas to which we have become accustomed from the study of the macroscopic world which surrounds us. For certain operations which are possible in the normal description of nature lose all significance in atomic physics because there they become in principle impossible.

For example, in normal life if we want to measure the position of a body we illuminate it in order that we can see it or photograph it; but if I want to establish the position of an electron I cannot behave in this way. It is impossible not for practical reasons but for conceptual reasons because if I illuminate an electron it will be struck by a quantum of light. Its motion will be affected by this collision as has been shown in a celebrated experiment by the American, Crompton. After the collision the electron will move in a direction which cannot be foreseen and with a velocity, depending on the direction, whose maximum value depends on the energy of the incident quantum. Thus the position of the electron changes because of the very fact that we observe it. The bodies which we observe in ordinary daily life are not disturbed by the light which falls upon them because of their very large mass; on the other hand the electron, because of its exceedingly small mass, moves with a considerable velocity when it has been struck by a quantum of light.

This example shows how critically we must always examine our arguments and how dangerous it is to transfer concepts which are valid in one field of phenomena to another field different from that in which they originated. In 1927 Heisenberg published a paper entitled 'On the Intuitive Content of Kinematics and on the Mechanics of Quanta'. In this paper he put forward and discussed his celebrated Uncertainty Principle which is a logical consequence of the bases of his mechanics. This principle establishes the maximum precision which one can obtain from the simultaneous measurement of the position and of the velocity of a particle.

We must examine carefully what we mean by 'the maximum *theoretical* precision which can be obtained'. The possibility of a simultaneous experimental determination of the position and of the velocity of a particle is excluded not because of limitations of the experimental apparatus or of the skill of the experimenter but because there is a theoretical reason why this simultaneous measurement cannot be made, even 'by a superman who owns perfect experimental equipment' (d'Abro).

Thus it is not conceptually possible to measure simultaneously the position and the velocity of a particle and to obtain for both measurements a precision as great as one chooses. In fact, as we already know, an electron under observation (with the help of an optical instrument) receives a blow from the incident quantum. It will, therefore, acquire a certain velocity and the greater the energy of the colliding quantum the greater this velocity will be. Now we are able to determine the position of the particle with as much precision as we require. To 'see' it all that we need do is to choose a radiation of sufficiently short wave-length, X-rays instead of normal light or gamma-rays instead of X-rays; but radiation of shorter wave-length implies a radiation of greater frequency, that is, quanta of higher energy. As a result the velocity of the particle which is struck by the incident quantum has a greater range of allowed values of velocity. The knowledge of the value of the velocity is therefore subject to an uncertainty which becomes greater when we arrange to measure the position more precisely.

If, on the other hand, we imagine an experiment which allows us to measure the velocity of a particle we find that the greater the precision which the apparatus allows us to achieve in the measurement of velocity the greater is the uncertainty in the position (once the measurement has been made). If the uncertainty in one of the two magnitudes is very small the uncertainty in the other will be very great.

Heisenberg discovered that the product of the uncertainty interval in the position of the particle by the uncertainty interval in its velocity is equal to a constant value; and this value is the constant h divided by the value of the mass of the particle. The same relation, called the uncertainty relation, is in general valid for any two complementary magnitudes: position and velocity of a particle or of a photon, time and energy, electrical intensity and magnetic intensity of an electromagnetic field at the same point of space, etc.

This then is Heisenberg's Uncertainty Principle; its consequences, as we shall see, enter also into the domain of philosophy.

No exception has yet been found to the Uncertainty Principle; it has been quantitatively confirmed in every imaginable experiment. It does not conflict with results of experiments in the macroscopic world as might appear when one thinks for example that direct measurements supply simultaneously knowledge of the position and of the velocity of a planet. For, as we have seen, the product of the uncertainties of the two quantities—position and velocity—is for a given body of mass, m, equal to h/m; thus it becomes smaller as the mass under consideration becomes larger. Now in the phenomena which occur in the macroscopic world the masses involved are so large that the uncertainties become absolutely negligible; they are very much smaller than other uncertainties due to our methods of measurement; but in the description of atomic phenomena very small masses are involved and this is the reason why the uncertainty relation which is masked in the macroscopic world by errors of measurement becomes in the atomic field clearly recognizable. 'Thus it is the microscopic which is profoundly real . . . it is there that we must seek the last hiding places of reality which in the macroscopic are hidden under the errors of data and in the confused mass of statistical averages.' (De Broglie.)

In a world in which Planck's constant has the value of zero there would exist no quantum phenomena; the uncertainty relation would vanish and classical mechanics would be completely valid even for the description of phenomena of the atomic scale. This would certainly make a great simplification; but such a world would be completely different from the physical world in which we live and in it life itself would probably be impossible.

CAUSALITY AND UNCERTAINTY

When we remember the criticism which was made of the concept of measurement in the atomic world, the criticism which was the starting point of the new quantum mechanics, we recognize the collapse of one of the most characteristic features of our normal conception of the world and of our relationships with it.

When, for example, an astronomer describes the movement of the moon he describes it as a movement which happens in space and time independently of the fact that he is observing it; it would happen in the same way with the same characteristics even if no human eye had ever been lifted to look at the sky. On the other hand, as we have seen, each act of observation on an atomic particle produces a change. As Bohr said, 'Thus there vanishes the separation between the observed object and the observing subject', and the observing subject may not only be a man but can also be a photographic plate. Thus, there arises again the old philosophical problem of the objective existence of phenomena independently of our observation.

In the concept of observation of ordinary phenomena (that is in classical physics) the distinction which exists between the observed system and the measuring instrument implied the validity of a rigorous determinism; the evolution of every physical system could be represented by a continuous succession of events linked by causal relations; the fundamental example of the way in which classical mechanics satisfied the causality principle was supplied by Newton's theory of the movement of the planets, a theory which is one of the great successes of mechanics. If at a given instant one knows the position and the velocity of the planets one is able to calculate exactly their position and velocity at a successive instant. The future and also the past of the system is completely determined when one knows its present state.

But there exist certain branches of physics in which one uses the concept of probability. For example, in the study of the properties of gases which are made up of an enormous number of molecules which move in a random manner, Maxwell and then Boltzmann introduced the statistical method; but this did not mean that one thought that the processes which take place in the inside of a gas fail to obey a rigorous determinism. The movements and the collisions of the single molecules of the gas happen according to rigorous laws and could, at least in theory, be completely described by a man who knew exactly at a given instant the position and the velocity of the single molecules. It is only because of the enormous number of molecules which make up a gas and because of their disorder that this prediction would not, in practice, be possible; and so statistical considerations are introduced and these allow one to obtain results which although incomplete are adequate for our purposes.

In this case, therefore, the uncertainty in the result is due to having made a statistical combination of many single processes, each of which, however, is controlled by rigorous laws. The uncertainty, therefore, is due only to the insuperable practical difficulty in being able to know at a given moment the position and the velocity of all of the very large number of molecules which make up the gas. It is only an expression of our incomplete knowledge of the data.

On the other hand, the uncertainty relationships of Heisenberg are inevitable and fundamental; they are the expression of a fundamental characteristic of nature herself.

Classical physics can then be defined as *completely deterministic* in the sense that the future behaviour of a system is completely and inexorably determined when one knows certain data about it present state.

This situation was fundamentally changed by the development of the new quantum theory. In fact in the atomic field, in which the distinction between object observed and observing subject vanishes, each time that one makes an observation there arises in the observed object a new and unpredictable factor and there is now no longer the possibility of knowing exactly and simultaneously all the data on the present state of the observed object (for example, position and velocity of a particle); and it is just this possibility which makes up the determinism of classical physics. In atomic physics we cannot decide whether the present state of a particle fixes its future state; we cannot prove whether there exist or not rigorous relationships of cause and effect.

Faced with this impossibility of proving a principle of causality one can then adopt one of two attitudes.

One may restrict oneself to saying that as a result of the existence of quanta no observation can lead to a simultaneous knowledge of the position and the state of motion of a particle without however thereby denying the existence of a trajectory and of a well-defined motion of the particle. Alternatively one can admit that a particle does not in general have well-defined values of position, trajectory or velocity; at every instant the magnitudes which classical theory attributes to the particle (its co-ordinance, the components of its velocity, etc.) do not have well-defined values but only a series of possible values each of which would have a certain probability. Only in the instant in which an observation was made or a magnitude was measured would this magnitude acquire a precise value. This increase in our knowledge concerning this particular magnitude would be balanced by an increase in our ignorance concerning its complementary magnitudes (position–velocity, time–energy).

The first attitude was adopted by certain scientists and in particular by Einstein who always claimed that the present theory would turn out to be only the statistical aspect of a deeper representation which would establish the existence of an objective reality.

But most theoretical physicists today follow Born, Bohr and Heisenberg in accepting the second interpretation. Thus, for them, an observation (or a measurement) has a significance which is different from that which it would have had in classical physics, in which it is a simple verification of an existing state.

The significance attached to an observation by the second group would also be different to that attached to it by a physicist who adopted the first attitude; for him, the observation would introduce a disturbance and therefore modify the existing state. For the second group, however, the magnitudes attributed to particles (velocity, position, etc.) do not have in reality exact values but only probable values; it is the measurement itself which creates for the observed magnitudes a fixed value.

Modern physicists are orientated in this direction; they thus abandon the belief in an objective reality which is independent of our knowledge; they come close to what in a philosophical field is known as idealism. Perhaps only the future can say if this interpretation can be considered as final.

THE COMPLEMENTARITY PRINCIPLE

Before finishing this rapid tour of the field of quantum theory I would like to comment on another principle which was put forward by Bohr in 1928.

As we have seen, the validity of the Uncertainty Principle has led to the need for abandoning the classical theory of causality in the field of atomic phenomena. There is accepted in its place a principle which Bohr has proposed as the result of a subtle and profound investigation. It is the so-called Complementarity Principle which may be considered as a compromise between a rigorous causality and a complete determinism. This principle allows us, as we shall see, to explain certain apparent paradoxes in atomic physics and in particular the wave-particle dualism which is exhibited both by matter and by light (and in all electromagnetic radiations).

Being able to establish a relation between a given effect and a given cause—that is being able to verify whether a rigorous causal connection obtains—makes sense only if it is possible to observe the cause and the effect without disturbing in any way the event. Now as we know, in atomic physics this is not possible; we must make observations if we want to localize an event in space and time—that is, if we want to know where it occurs at a given moment; but a precise localization in space and time leads to an uncertainty in the development of the

Atomic Physics

future; and vice versa, a rigorous connection between cause and effect can be described only when we deny ourselves accurate location in space and time.

Bohr generalized these results in his Complementarity Principle. 'In atomic phenomena it is not possible to achieve simultaneously a rigorous localization in space and in time and a rigorous causal description.' Space-time localization and causal description are complementary. They represent two different aspects of reality which cannot both be rigorously and simultaneously defined.

One should note carefully that it is not impossible that atomic processes should be connected by strict relations of cause and effect; but it is impossible to define accurately in space and time a series of events which are causally related. It is in this sense that we have stated that the Complementarity Principle does not completely reject causality but is rather a compromise between rigorous causality and complete free will.

And as usual this appears to be valid only for atomic phenomena and not for the usual phenomena of the macroscopic world because only in the first case is the value of the constant h not negligible compared to the magnitudes involved.

OVERCOMING THE WAVE-PARTICLE PARADOX

Bohr showed how this complementarity appears in questions of the nature of light and the nature of matter.

We saw that, following on the discovery of the photo-electric effect and on the development of wave mechanics, the conclusion was reached that light and matter show a dual nature; sometimes they behave like waves, sometimes like particles; and we asked ourselves: how can this be possible?

If one clings to classical concepts one is confronted by an insoluble dilemma but in the light of the complementarity principle this well-known dilemma disappears because it does not exist. The properties connected with the wave nature (of light or of matter) and those connected with their corpuscular nature are complementary; they can never exhibit themselves together in the same experiment because this is forbidden by the Uncertainty Principle. When one conducts an experiment which exhibits the wave properties (for example a diffraction experiment) this experiment produces a disturbance which makes the corpuscular nature vague and unrecognizable. When one makes an experiment which reveals the corpuscular nature the properties connected with the wave nature become vague and unrecognizable.

'The wave and corpuscular properties', wrote de Broglie, 'never come into conflict because they never exist simultaneously. One continually anticipates the duel between the wave and particle; but it never happens because only one of the two rivals is present.'

The corpuscular aspect and the wave aspect are not contradictory but are two different aspects of the same reality; the impossibility of revealing them both simultaneously, expressed by Heisenberg's uncertainty relations, is due to the existence of Planck's constant of action.

Our intuition does not allow us to make a complete representation of this reality. Inspired by common experience we construct simple models, models which are too simple, by means of which we try to interpret phenomena: the corpuscular model and the wave model; but we now know that each of these models or, as Bohr has called them, each of these idealizations is by itself inadequate to describe the complexity of reality; according to the circumstances we must use either one or the other of these idealizations despite their mutually contradictory character.

A physicist has said, 'the more or less schematic idealization in our minds can represent certain aspects of things but these idealizations are limited and cannot contain in their rigid frames all the richness of reality'.

CONCLUSION

Thus we have arrived at the end of this history of atomic theories. Democritus presented to his contemporaries a world made up of indivisible, unalterable atoms in eternal motion. Today, physicists offer us a mathematical framework which allows us to deal with all the phenomena which occur in the atom, excluding always the complex structure of the nucleus.

All the models which have been proposed for the constituents of the universe—radiation and matter—have as time went on shown themselves to be inadequate; we can use one or the other according to the particular problems to be treated but we must not forget that, as I have said in the Introduction, nature is perhaps more complex than our intuition allows us to understand.

CHAPTER FOUR

NATURAL RADIOACTIVITY

We have seen that in the study of atomic physics the nucleus is treated as a single particle of negligible dimensions and is characterized only by the values of its mass and of its positive electric charge.

But this description is certainly incomplete. The first demonstration of this fact is provided by radioactive phenomena which consist in the spontaneous disintegration of the atomic nuclei of some heavy elements. From this we see that an atomic nucleus must be made up of smaller particles.

It is logical therefore that before discussing the structure of the nucleus (its constituent particles and the laws which they obey) I should consider further the phenomenon of radioactivity, show how it appears in nature and describe the methods which are used to detect the particles and the radiations emitted in radioactive disintegration.

ISOTOPES

In the course of the enormous work carried out after the discovery of radioactivity one fact emerged which appeared very strange. It was discovered as a result of the use of chemical methods to examine the chemical properties of the elements which are formed in the disintegration of various radioactive atoms. It was realized that in certain cases the chemical properties of two elements (which were known to be two different elements) were so similar that if they were mixed together it was impossible to separate them by any chemical method. Now, as we know, the chemical behaviour of an element (and to a large extent also its physical behaviour) depends on the number of planetary electrons; so if there exists in nature two different elements which are chemicaly indistinguishable their atoms must possess the same number of external electrons. They must then occupy the same place in the periodic table of elements; for we have seen that the number, Z, of an element in Mendeleev's table (its atomic number) is just equal to the number of planetary electrons of the atom of that element. But the atoms of different elements differ one from another both in the number of the external electrons (that is in the value of the charge of the nucleus) and in the weight of the nucleus. Thus two separate elements which have the same number of electrons can differ only in the weight of their atomic nuclei.

When, therefore, in the study of radioactivity it was found that in certain cases it was not possible to separate by chemical means two elements (which were, moreover, very different in their radioactive properties) Soddy concluded that these must be elements of different atomic weight which had, however, the same atomic number, Z. As these elements occupied the same place in Mendeleev's periodic table of elements (having the same atomic number) Soddy gave them the name of *isotopes* (from the Greek isos topos meaning 'equal place'). For example, there exist two elements which have the atomic number 17; that is in Mendeleev's table both occupy the same box number 17 (chlorine). These elements which are chemically identical differ in their atomic weights which have respectively the values 35 and 37 (taking as usual as the unit one-sixteenth of the weight of the atom of oxygen, i.e. virtually taking the weight of the nucleus of hydrogen as equal to one). The chlorine which is found in nature is a mixture of these two isotopes; but whilst the isotope with the atomic weight 35 makes up 75·5% of natural chlorine, that of atomic weight 37 contributes 22·5%. This is why chemists found that the atomic weight of chlorine was equal to 35·45.

An impressive series of experiments was carried out by F. W. Aston and others, using an apparatus called a *mass spectrograph* which I will describe later. As a result of these experiments, it was discovered that most of the chemical elements are in reality a mixture of isotopes. There exist in nature only 92 chemically distinct elements, but the number of different atoms (different in weight or in nuclear charge) is very much larger; it is about 240.

Not all isotopes are stable. Some are formed, as we shall see, in the disintegration of radioactive elements and they in their turn sooner or later disintegrate.

The discovery of the isotopes led to a fundamental investigation which, as we shall see later, forms the basis on which the modern theory of the atomic nucleus was built. About a century ago the chemist, Prout, had made the hypothesis that the atoms of different elements were all made up of atoms of hydrogen; that is, that the simplest of the atoms was the sole material with which the atoms of all the other elements were built. But when measurements showed that, taking as a unit the weight of the hydrogen atom, the atomic weights of the ninety-two elements in Mendeleev's table are not, apart from a few rare exceptions, whole numbers, this hypothesis had to be abandoned. Thus, to the great disappointment of almost all scientists from Mendeleev on, it was not possible in any way to attempt the construction of a theory in which the separate elements would comprise assemblies of various numbers of a single, or of at least, a few types of particles.

The hope of being able to make such a theory was resuscitated as soon as it was seen that, although the atoms of the ninety-two elements had weights which could not be represented by integers, the weights of the 240 known isotopes were represented by numbers which were very nearly integral. Now this is just what should happen if the atomic nuclei of all elements were made up of particles all having the mass of the nucleus of hydrogen. We will investigate later the validity of this hypothesis in modern nuclear physics and see the form in which Prout's hypothesis has been re-established.

The nucleus of the hydrogen atom acquires therefore a particular importance. It is the unit of atomic mass and the unit of positive electric charge. Because of its particular importance it deserves a special name; and it was called *proton* (from the Greek protos = first).

RADIOACTIVE TRANSFORMATIONS

After 1900, researches on radioactivity brought to knowledge a complicated assembly of facts. Many radioactive substances were discovered. It was discovered that some of these lose their radioactivity within a few hours or in a few days whilst others do not appear to show any diminution of activity with time. This collection of facts appeared at first to defy every attempt at interpretation; but it was put in order and logically interpreted in 1903 by the theory of radioactive transformations put forward by Rutherford and Soddy. 'According to this theory', wrote Rutherford himself, 'the atoms of radioactive elements, in contrast with the atoms of ordinary elements, are not stable but disintegrate spontaneously emitting an alpha particle or beta particle. After the disintegration the residual atom has physical and chemical properties which are completely different from those of its parent atom. It may, in its turn, be unstable and pass through a succession of transformations, each of which is characterized by the emission of an alpha particle or of a beta particle.' I should add that frequently each of these disintegrations is accompanied by the emission of energy in the form of a gamma ray.

In order to explain how an atom is modified after the expulsion of a particle we must distinguish between two cases, according to whether the particle emitted is an alpha particle or a beta particle.

By the emission of an alpha particle, that is a helium nucleus (atomic no. 2, atomic weight 4) the weight of a radioactive atom is reduced by four units. The emission of a beta particle, on the other hand, hardly changes at all the weight of the atom since an electron has a mass which is 1,800 times smaller than the mass of the lightest atom, hydrogen.

Just as important as the change in the mass is the change in electric charge. An alpha particle has a positive charge plus 2, and an electron has a negative charge minus 1; so when a radioactive nucleus emits an alpha particle its charge is reduced by two units, and when it emits a beta particle, it loses a charge minus 1, that is, its positive charge is increased by one unit.

To sum up, then, 'if a radioactive nucleus emits an alpha particle, its weight is reduced by four and its charge by two; if it emits a beta particle its weight remains almost unchanged and its charge is increased by one'. This is the so-called *displacement law*.

We know that in the periodic system the elements are arranged by increasing nuclear charge and that the ordinal number of each box coincides with the nuclear charge of the elements which occupy it. Thus the loss of two positive charges implies the displacement by two places in the periodic table towards the boxes with lower atomic numbers. The loss of a negative charge, on the other hand, corresponds to a displacement of one place towards the boxes with higher atomic numbers.

Let us look at a particular example, and see what happens in the disintegration of radium, which is an alkali earth element of atomic number 88, and atomic weight 226. Its atoms are not stable but have a certain probability of disintegrating; we shall see later how to interpret this probability. Let us suppose that we can choose a single atom which is about to disintegrate. It does so, 'with the violence of an explosion' and expels an alpha particle (atomic number 4, charge $+ 2$). The atom which is formed after the reorganization of the residual nucleus must have an atomic weight four units less than that of radium ($226 - 4 = 222$) and must lie in the periodic system of elements displaced two places towards the lower atomic numbers. As radium has atomic number 88 one of its atoms must pass on disintegration to the box 86 which is in the column of noble gases. One finds in fact that the atom of radium, in agreement with what one would expect from these theoretical considerations, changes into an atom of emanation of radium or radon, which is the heaviest of the noble gases. The fact that radon is a gas is alone sufficient to show that it has chemical properties very different from that of radium.

The atoms of radon themselves are not stable; they too have a certain probability of disintegrating and when they do so they too emit an alpha particle; an atom of radon is changed by the loss of an alpha particle into an atom of an element which is called radium A (RaA) of atomic weight $222 - 4 = 218$ and atomic number $86 - 2 = 84$; it is in its turn, radioactive.

We will not, at present, follow the further transformations which lead from the unstable atom of radium to a stable atom which is, as we shall see, an atom of lead. I have summarized the transformations from radium to radium C′ in a table (Fig. 11); one sees immediately that both radium C′ and radium A have charge 84 and are therefore isotopes; their atomic weights differ by four units.

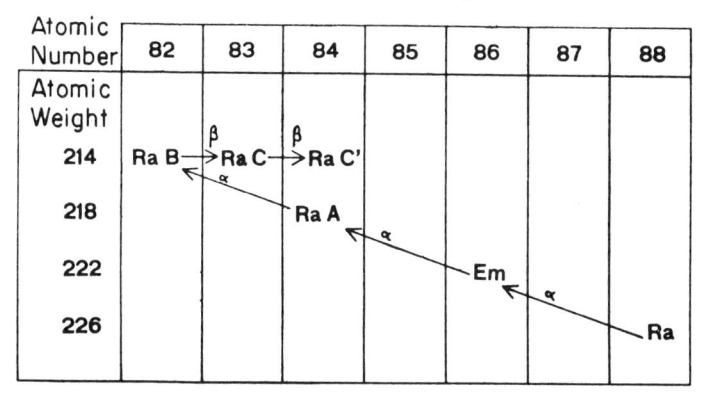

Atomic Number	82	83	84	85	86	87	88
Atomic Weight							
214	Ra B $\xrightarrow{\beta}$	Ra C $\xrightarrow{\beta}$	Ra C'				
218			Ra A				
222					Em		
226							Ra

FIG. 11. Radioactive transformation from radium into radium-C'

THE RADIOACTIVE FAMILIES

The terms which are used in the study of radioactive transformations are particularly easy because they are derived from analogy with human life.

Radioactive substances are generated one from another; thus radon is generated from radium and radium A from radon, etc. The formation of an atom of radon is considered as the death of an atom of radium and the birth of one of radon; and one says that radium is the mother substance of radon. Just as for mankind, one calls the life of an atom the time which lapses between its birth and its death.

The analogy can be pushed further because radioactive substances again like mankind are organized in families. All the heavy radioactive substances are derived originally from uranium and from thorium the heads of three different families called respectively uranium-radium, uranium-actinium, and thorium (Fig. 12).

The family of uranium-radium is that which contains the greatest number of components: to it belong, in addition to the head which is uranium and other members (uranium X_1, uranium X_2, uranium II, etc.), also ionium, radium and emanation of radium which, in contrast to the other members, is a gas.

Each of the two families of thorium and uranium-actinium also possess a gaseous component which is called respectively

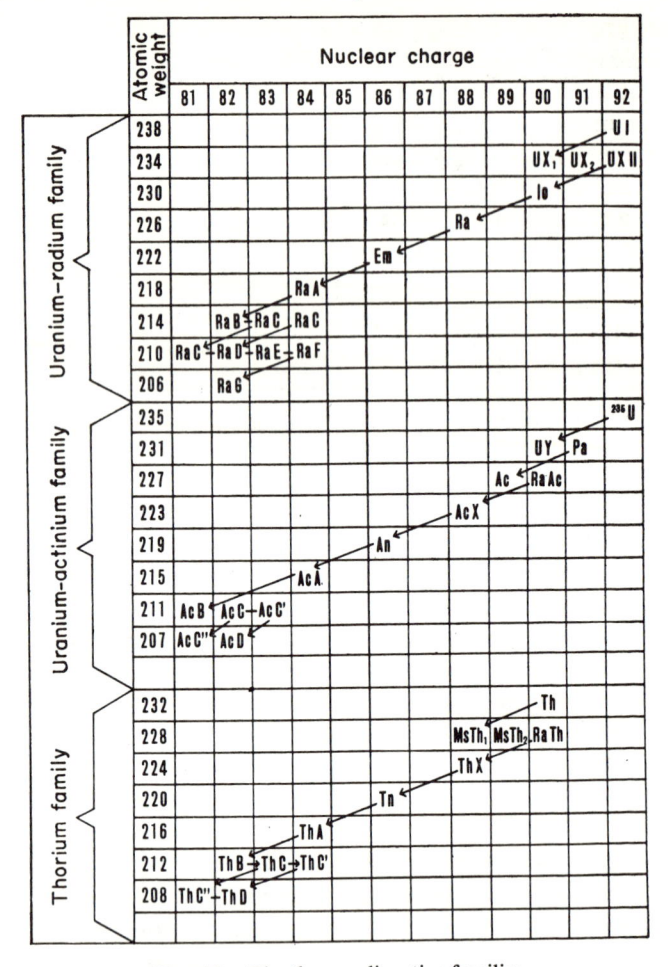

FIG. 12. The three radioactive families.

emanation of thorium and emanation of actinium. It is interesting to note that the three emanations are isotopes; for all three have the same charge. Their atomic weights depend on the transformation type (alpha or beta) which their ancestors have undergone and on the mass of the head of the family.

The final stable products of all three of these radioactive families are isotopes of lead.

In addition to the radioactive substances which belong to

these three families there exist in nature some others which do not belong to the group of heavy elements. These are: the isotope of atomic weight 152 of samarium which has a life of some hundreds of millions of years (10^{11} years); the isotope of atomic weight 40 of potassium which has, it appears, a life of the order of 100 million years and which yields amongst its disintegration products argon of atomic weight 40 which is present in the atmosphere; and then rubidium and, very weakly, lutetium.

THE LAWS OF RADIOACTIVE DISINTEGRATION

It has been found experimentally that the law according to which radioactive atoms break up is exceedingly simple: the number of atoms of a radioactive substance which disintegrate in one second is proportional to the number of atoms present. This means that, for example, if out of one hundred atoms of a given radioactive substance three disintegrate in one second then in a sample of the same substance containing 100,000 atoms 3,000 will disintegrate in a second. The choice of the atoms which break up is left entirely to chance, and as Soddy has said, one could imagine a destructive angel which seizes here and there at random the atoms which must break up, maintaining, however, a fixed proportion between the existing atoms and those which must be destroyed.

This law of disintegration is the same for all radioactive substances. On the other hand there is enormous variation from one radioactive substance to another in the rhythm with which each disintegrates, that is in the amount which disintegrates in a given time.

One calls the half-life or the period of a radioactive substance the time in which half of a quantity of that substance will have disintegrated; that is the time needed for a certain quantity of that substance to be reduced to a half. Now, this time varies for different radioactive substances—from some thousands of millions of years to a minute fraction of a second. For example, thorium has a half-life of 14,000,000,000 years, radium of 1,590 years, and thorium C of only one thousandth millionth of a second. The heads of the three families are characterized by exceedingly long half-lives which can be compared in order of magnitude with the age of the earth. This fact explains how it is that radioactive substances can exist. If, for example, uranium were to remain alive for only a few years it would by now have completely disappeared and with it there would also disappear the members of its family; but it disintegrates very slowly, continuously regenerating the whole series of its successors and in the same way thorium and protactinium disintegrate extremely slowly.

This disintegration rhythm, which for every radioactive substance is constant, is not only independent of any physical or chemical condition but is also independent of the age of the substance. An atom of radium has today a certain probability of disintegration; but if after a thousand years it has not yet disintegrated it will have on that day the same probability of disintegrating in twenty-four hours that it has today. Nobody knows for how long a radioactive atom will live; it may disintegrate in the very next moment or it may continue to live for centuries, but one thing is certain: if we have a certain number of atoms of this element they will, after a time equal to the half-life, be on average reduced to a half.

An atom has always the same probability to disintegrate whatever may be its past history or its present state. As Jeans has said, 'Here there was a law of nature of a type until that moment unknown to science and its consequences, as one can immediately see, were enormous. From Democritus through Newton up to the nineteenth century science had proclaimed that the present was determined by the past; the new science on the other hand seems to be saying something very different and that in the events now under consideration the past does not seem to have had any influence on the present nor the present on the future.'

For the first time man found himself confronted with a phenomenon which did not agree with his classical method of thinking according to the law of causality; for the first time there entered into science in this sense the concept of probability.

Naturally one could think that there exists some hidden cause, until now undiscovered by man, for which an atom lives only for a few seconds whilst its neighbours, which are apparently identical and which exist in the same conditions, survive for several years; but this would be a speculation which has no experimental foundation; indeed, as we have seen, the new quantum mechanics states that it is senseless even to look for this cause; and I have already shown how profound has been the evolution in atomic physics of the concept of causality.

RADIOACTIVE EQUILIBRIUM

The quantity of any radioactive substance will then decrease with time and be reduced to a half after a fixed period. This is what happens when one considers a radioactive substance separated from another substance from which it arises; when, on the other hand, there is a source which regenerates the substance things naturally work out differently.

Let us choose as an example the case of the transformation of radium into emanation. Let us suppose then that we observe a certain amount of radium after we have removed all the decomposition products (radon, radium A, radium B, etc.). The radium will go on disintegrating to produce radon; and radon in its turn will be transformed into radium A; but not all of the radon which is formed disintegrates immediately. Thus we will find at the beginning of our observation that the amount of radon will increase with time. However, as it increases there increases also the number of its atoms which break up. At a certain time we will reach the state when there are as many atoms of radon formed from radium as there are atoms which decay and are changed into radium A. From this time on the amount of radon will remain constant. It is then said that 'the radium and the radon are in radioactive equilibrium'.

The same thing happens in the transformation of radon into radium A (in the absence of radium). Certain atoms of radon are changed into atoms of radium A; of these only a small percentage breaks up and thus initially the amount of radium A will increase. As it increases so will the number of its atoms which break up and thus, after a given time, the amount of radium A which breaks up will be equal to the amount which is created.

There is, however, a difference between the first case (the transformation of radium into radon) and the second (the transformation of radon into radium A in the absence of radium); this difference arises from the different mean lives of radium and of radon. Radium has an extremely long mean life and is reduced to a half in about sixteen thousand years. Thus the percentage of radium which breaks up even in a fairly long interval is so small that the original amount of radium can in practice be considered as constant; as a result the number of its atoms which break up every second is also constant. Thus once equilibrium has been reached the radium will continually regenerate radon which disintegrates and the amount of radon in equilibrium with the radium will remain constant.

But in the second case, on the other hand, the radon which is reduced to a half in less than four days tends to disappear fairly quickly. It is true that one reaches an equilibrium but his equilibrium is ephemeral. Very soon the radon does not supply to radium A as many new atoms as there are atoms of radium A which disintegrate; and therefore the amount of radium A tends to decrease.

The time needed for a radioactive substance to reach equilibrium with the mother substance is fixed for every transformation; as one can easily see it depends on the rate at which the substances break up. Thus radon reaches equilibrium with radium about fifteen days after the start of the process of generation; on the other hand, radium A reaches equilibrium with radon in about twelve minutes.

All the transformation products of a given substance reach radioactive equilibrium so that when the head of a radioactive family is not separated from its decomposition products the latter will reach radioactive equilibrium. And as the life of the three heads of family is extremely long their activity can, in practice, be taken as constant. Thus we should expect to find that the proportions of different members of the family should be constant when they are in the presence of the head of the family. In fact, if one measures, for example, the amount of radium present in various types of uranium bearing rock one finds that the percentage of radium and of uranium are always the same. An important example is that of stable lead. There is a constant amount formed per unit time from the successive disintegrations of the elements belonging to the three radioactive families and as the lead is stable it accumulates in the rocks; as we shall see, a measurement of the relative amounts of lead and uranium contained in a rock can be used to establish its age.

RADIOACTIVE SUBSTANCES IN NATURE

When in 1903 Pierre and Marie Curie were looking for radioactive substances and discovered radium they had to examine about seven tons of pitch-blende coming from the deposits at Joachimstal in Bohemia in order to prepare a single gram of radium bromide. They had chosen a material which was particularly rich in radioactive substances; if they had, on the other hand, taken a material at random they would have had to handle a quantity a hundred thousand times greater in order to obtain the same gram of radium bromide.

Very small quantities of radioactive substances are found nearly everywhere. Without going too far afield one can find traces in, amongst other things, building materials. The walls of a medium-sized building may contain about fifty kilogrammes of radioactive substances. Most of these are, however, uranium and other elements which have a very long mean life; these disintegrate so slowly that they can be considered as almost stable. The amount of substances with short mean life, on the other hand, is very small; in the walls considered above, for example, there might be about five milligrammes of radium, and if one were able to separate all the radium contained in the huge group of Mont Blanc one might be able to fill a wheelbarrow.

Radioactive substances are also contained in the sea and in the great rivers but in a percentage about a thousand times less than that of the rocks. There are, however, some natural sources of water which are very much more radioactive than that of the sea and the rivers.

THE AGE OF THE ROCKS

A problem which has intrigued geophysicists for a very long time is the determination of the age of the rocks or more correctly of the epoch in which the various rocks reached the solid state in which we find them today. The spontaneous disintegration of radioactive elements has supplied the means for tackling this problem.

This disintegration is in fact a regular phenomenon like the movement of a natural chronomoter, always precise and completely unaffected by external influences such as the temperature and the pressure. It occurs in time at a well-defined rate so that it is possible to establish the history of a given amount of material. One can, for example, determine what percentage of a gram of uranium will still exist after a thousand years, or for a single gram of uranium which exists today determine the amount which existed a million years ago.

The method of calculating the age of a uranium bearing rock is based on the fact that uranium as we already know has a mean life of the order of some thousands of years, and is transformed via all the members of its family into an isotope of lead; and furthermore the time required for a given quantity of uranium to be transformed into a given quantity of lead is known exactly, and thus from the measurements of the percentage of uranium and of this isotope of lead which are found together one can calculate the time which was required to make the transformation.

It has been found by very laborious calculations that the oldest rocks are about two thousand million years old; and it has been found that the oldest meteorites are of about the same age.

Thus, in the history of the earth radioactive substances have acted as a very precise clock which was wound up in the course of the formation of the rocks and which has continued to tick regularly until today.

THE INTERNAL HEAT OF THE EARTH

Knowing the amount of radioactive products which are found in the earth's crust we have been able to tackle the very old problem of the origin of the heat inside the earth.

It is well known that the temperature increases as one descends underground; on average it goes up by about one degree centigrade in every thirty-five metres (*the thermal gradient*). This implies that there must arise from the depths of the earth a continuous flow of heat which is lost to the outside. What is the origin of this heat which from the start of life on earth has escaped from its centre?

Natural Radioactivity

It is not a negligible amount of heat since it amounts for the whole surface of the earth to six billion calories[1] per second; this means that in a million years forty million calories have escaped from each square centimetre of the surface of the earth. Kelvin and others supposed that this continuous flux of heat was a 'souvenir' of the time in which our planet was fluid and had a high temperature. According to this hypothesis, as Kelvin discovered by a simple calculation, the current value of the thermal gradient should have been reached only thirty million years after the solidification of the earth's crust; but it is now known that the earth has existed for about four thousand million years. Thus, Kelvin's hypothesis is completely invalid.

It appears certain that not more than twenty per cent of the observed heat flux can be due to the primeval heat which existed in the earth when it was formed. The rest must have some other origin. What is it? Thanks to the discovery of radioactivity this problem has been solved.

In 1903 Pierre Curie and Laborde discovered that radium salts continually yield heat: one gram of radium emits rather more than two calories a minute and in the course of its whole life develops three thousand million calories. If one wanted to get these three thousand million calories by burning coal one would need more than half a ton.

All radioactive substances produce heat; but the amounts naturally differ from substance to substance. For example, a year would be required to obtain one calorie from forty kilogrammes of potassium whilst the same amount of heat could be obtained from one gram of radium in thirty seconds.

To understand the production of heat by radioactive substances we must remember that they emit particles (alpha and beta) and gamma rays. Now as the particles have a mass and a velocity they possess kinetic energy. If the mass of the particles is very small their velocity is very big. In the case of beta particles the velocity may be very near to the velocity of light. Thus the energy of a single particle is certainly not negligible; and given the huge number of particles emitted by radioactive substances the total energy of the alpha and beta rays is very large; to this we have to add the energy of the gamma rays

---

[1] One calorie is the amount of heat which is needed to raise one gram of distilled water from 14·5 to 15·5 degrees centigrade. One billion equals 10^{12}.

which is also not negligible. Since every form of energy always ends as heat radioactive substances must eventually produce heat.

Thus the radioactive substances which are found in the earth are responsible for a large part of the heat which continually escapes from it into outer space, but there now arises a new difficulty: Kelvin's theory supplied an amount of heat which was inadequate to account for the actual geothermal gradient; the radioactive theory on the other hand supplies too large an amount of heat. In fact, if one assumes for the whole of the earth a radioactive level analogous to that found in the earth's crust the heat which would be produced by all the radioactive substances would be very much greater than that which the earth loses by radiation. Thus our planet would heat up rapidly.

As this does not happen one must assume that the percentage of radioactive substances decreases as one moves from the crust towards the centre of the earth. An explanation of this decrease has been proposed by Joly and Poole. Let us make the hypothesis that our planet was initially fluid and of a homogeneous composition. Gravitation should then, with time, bring together at the centre the heaviest constituents and leave at the surface the lightest; but radioactive substances which, because of their weight, should accumulate around the centre liberate heat and therefore increase the temperature of the materials in which they are contained. The latter, therefore, behave like materials of lower density and are grouped with the lighter substances which make up the superficial rocks.

It is found in support of this hypothesis that meteorites—these samples of other planets which reach our earth—do not all contain the same proportion of radioactive substances. The meteorites which contain mainly iron, that is the heavier meteorites, are less radioactive than the lighter ones whose composition is similar to that of the earth's crust.

COSMIC RAYS

Before closing this chapter I would like to refer to another indirect contribution to science made by radioactivity.

It has been known for some time that the atmosphere of the earth is lightly ionized. After the discovery of Becquerel it was

Natural Radioactivity

thought that this ionization could be due to the passage into the atmosphere of radiations emitted by terrestrial radioactive substances. Gamma rays would make up the main contribution as they are the most penetrating. There would also be small quantities of radioactive substances such as an emanation of radium present in the atmosphere. If the ionization of the atmosphere were really due to these radiations it should diminish with height and new measurements were therefore made at various heights; but the results conflicted with the predictions. Not only is the atmosphere ionized at very high levels which could not be reached by the most penetrating radiation produced in the earth; but also above a certain level the ionization increases with height. Thus the existence of a new type of radiation was recognized. As this radiation is very much more penetrating than all others yet known it has been called penetrating radiation and as these rays come from cosmic space they are also called *cosmic rays*. I shall discuss these in greater detail in a later chapter.

CHAPTER FIVE

HOW RADIOACTIVE PARTICLES ARE DETECTED

THE PASSAGE OF CHARGED PARTICLES THROUGH MATTER

An atomic nucleus which breaks up can emit, as we already know, three different types of radiation: alpha particles (helium nuclei), beta particles (electrons) and gamma rays. What are the methods which are used to detect these radiations? What is it that enables us to recognize their nature, to measure their energy, to observe their trajectories, and in the case of corpuscular radiation to determine the mass of the particles, the sign, and magnitude of their electric charge and finally to count them.

Naturally, because of their very small size it is not possible to see alpha particles or electrons, even using the most powerful microscope; one must therefore be satisfied with following them indirectly. We do not see the wind but we can follow it by observing the effects which it produces: it bends the tops of trees, it raises dust, disturbs the sea, whistles ...; by observing these effects we can determine its intensity and its direction. Analogously, although we cannot directly observe radioactive radiations we can study the effects which they produce in their journey through material.

When a charged particle possesses adequate energy it can pass through matter and in particular can pass through a gas and in doing so collides with some of the constituent atoms. An atom is a relatively large target for radioactive particles; but almost the whole of this target is occupied by the cloud of planetary electrons whilst the nucleus is only a minute point at the centre. Thus the collisions of a charged particle against an atom are effectively collisions against the electrons; colli-

How Radioactive Particles are Detected

sions with the nucleus—which is at the centre and well protected by the electron cloud—are so rare that we can for the moment neglect them.

The electrons themselves are extremely small and it might at first seem surprising that the collisions occur so frequently. The two particles which are charged with electricity exert forces on each other even at a distance. They are electrical forces: they are attractive if the two particles have electric charge of opposite sign and repulsive if the charges are of the same sign and so in collisions with charged particles an electron behaves as if it were very much larger than it really is.

During the collision the charged particle gives to the electron some of its energy. The electron which now has a higher energy jumps to revolve in a more external orbit corresponding to this new energy. The larger the amount of energy which the electron acquires from the collision the farther from the nucleus will the new orbit be. The atom is then 'excited'. After a very short time the electron will return to its original orbit and will emit in the form of light the energy which it received from the colliding particle. If the energy transferred by the colliding particle is so large that after the collision the electron has an energy which is greater than the energy with which it is bound to the nucleus it will be torn out of the atom. The atom is then left with a positive electric charge (i.e. it is an ion and the atom is ionized).

And so when it passes through matter a charged particle leaves a trail of excited atoms, electrons and positive ions; this trail is the result of the passage of the particle. By observing the trail—and we shall soon see how we can do this—one can acquire knowledge of the properties of the particle which produced it.

The charged particle as it moves through matter dissipates most of its energy in this way until it loses its power of further penetration. It can be shown that for different particles which all have the same initial energy and which cross the same amount of material the number of ionized atoms along the path is the greater the greater the mass of the particle. Thus an alpha particle should produce a greater number of ions than an electron with the same initial energy. For example, an alpha particle from polonium yields about forty thousand ion pairs in one centimetre of its path through air which is at normal temperature and pressure; and since in every collision it gives up some of its energy it will slow down until it finally stops after a few centimetres. In the same air an electron of the same energy would produce only about fifty ion pairs per centimetre of path and thus its penetrating power is very much greater than that of the alpha particle.

THE PASSAGE OF PHOTONS IN MATTER

As we now know, gamma rays are made up of electromagnetic radiation of the same type as light waves, ultra-violet rays and X-rays; they have the highest frequency of all electromagnetic radiations.

Light, ultra-violet rays and X-rays are made up of quanta which are emitted when one of the planetary electrons of an atom jumps from one orbit to another lower orbit. The emission of gamma rays is rather more complicated because they come from the nucleus; gamma radiation is also made up of photons which are emitted when a nucleus passes from one quantum state to another which has a lower energy.

Let us now see what happens when a high energy photon crosses matter. It loses energy and generates ion pairs by means of three fundamental processes: the photo-electric effect, the Compton effect and materialization. One or other of these processes will occur according to the energy of the photon and to the atomic weight of the substance which is penetrated.

If the photon which collides with an atom has an energy which is large enough to ionize it one of the planetary electrons of the atom will be torn out; this electron will move away from the nucleus with an energy which is equal to the difference between the energy of the photon and the energy which is required to extract the electron from the atom. This process consists then in the absorption of a photon by an atom associated with the expulsion of an electron; it is called the photo-electric effect.

If the colliding photon has a much higher energy—an energy which is much greater than the energy which binds an electron to the nucleus—the binding energy can be neglected; the electron then behaves as if it were free and not bound to the nucleus. Thus the process is similar to that of an elastic collision between a photon and a free electron. The same thing happens as that which occurs in a game of billiards when a moving ball strikes one that is stationary. After the collision the first ball will move in a new direction with a lower velocity whilst the ball which was originally stationary will acquire a certain velocity. The moving ball has lost energy to the ball which it has set in motion. Analogously the photon, when it collides with an electron (Fig. 13) is deflected and loses some of its energy which is acquired by the electron which it sets in motion. The photon which emerges has therefore less energy than the incident photon.

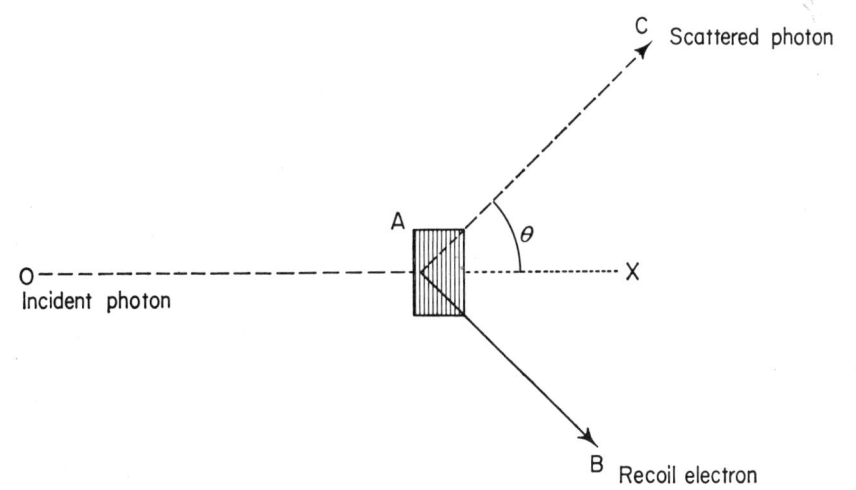

FIG. 13. Compton effect.

And remembering that the energy E of a photon is related to its frequency by the equation $E = h\nu$ we see that the scattered photon will have a lower frequency than the incident photon.

This is just the Compton effect which the American physicist A. H. Compton discovered in 1923 and which carries his name. He observed that when a beam of X-rays crosses matter part of it is scattered in all directions (as happens when light crosses a turbid medium) and that the frequency of this scattered radiation is rather less than that of the incident radiation. The Compton effect could not be explained in terms of classical theory. It provides a brilliant experimental confirmation of the hypothesis of light quanta.

The lighter the material in which the collisions occur for a given energy of incident quanta the more pronounced is the Compton effect; for the atomic nuclei of the lighter elements have small electric charges and therefore in their atoms the electrons are less strongly bound to the nucleus; nearly all of the electrons therefore behave as if they were free. One can also see that for a given substance the Compton effect will be more apparent the higher the energy of the colliding photon, that is the higher the frequency of the incident radiation.

Thus electromagnetic radiation which crosses matter can produce in it either the photo-electric effect or the Compton effect; and it will produce one or the other according to the energy of the incident photon (that is the frequency of the radiation) and the weight (that is the atomic number) of the material which it crossed.

Finally the third process by means of which a photon, when it crosses matter, can lose energy is materialization. Later I shall return to this interesting phenomenon which can occur only when the incident photon has an exceedingly high frequency (corresponding to a wave-length of about one ten thousandth millionth of a centimetre); it is found that this phenomenon is effectively possible only when the photon passes through the electric field of a nucleus.

METHODS OF DETECTING THE PARTICLES

Thus both charged particles and electromagnetic radiation produce along their paths excitation and ionization of the atoms of the material they encounter. This excitation and this ionization form just those effects which we are seeking; they correspond to the bending of the trees, to the raising of dust and to the waves produced by the wind. There are various methods to make visible or to measure these effects; they can be divided into two classes according to the information which they supply.

To the first class belong those instruments (which are called by the general name of counters) which give a signal when they are crossed by a particle. They allow one to localize with very great precision the time when the crossing occurred but they give only vague information about where it occurred; that is they localize very well the trajectory of a particle in time but not in space.

The detectors of the second category (which I will call with the general name of track detectors) allow on the other hand good localization of the particle in space but only poor localization in time.

Using all these instruments physicists tried to determine the trajectory, the mass, the velocity and the electric charge of these particles which escape from direct observation.

COUNTERS

We have seen that a charged particle produces excitation and ionization of the atoms which it encounters in its path; this is the microscopic signal given by the passage of a particle. It is a signal which we cannot perceive with our senses; it must be transformed.

Various types of counters have been thought of, some of which exploit the ionization produced by the particle and others, on the other hand, the excitation. The first are the ionization chamber and the counter as such; and the second are the scintillation counter and the Cerenkov counter.

The Ionization Chamber

When the particle crosses a gas it generates in the gas electrons (that is negative electric charges) and ions (that is, positive charges) and naturally the greater the number of broken atoms the greater the number of these charges. One may therefore obtain a measurement of the ionization produced by the passage of a particle in the gas by collecting the charges which it generates. This is just what the ionization chamber does (Fig. 14). It consists essentially of a gas-tight vessel containing gas and two metal plates (electrodes); one of these is charged positively and the other negatively.

If particles are passed through the gas it becomes ionized and therefore a certain number of electrons and positive ions are produced. The electrons which have negative electric charge become attracted to the positive plate and the ions which are positive to the negative plate. The pressure of the gas and the charge on the electrodes can be regulated so that all the charged particles which are produced will arrive on the electrodes.

The assembly of positive ions which are discharged on the negative electrode produce a small reduction in its charge whilst the electrons produce a small reduction in the charge of the positive plate, thus producing a reduction in the difference of potential between the two plates. The greater the number of

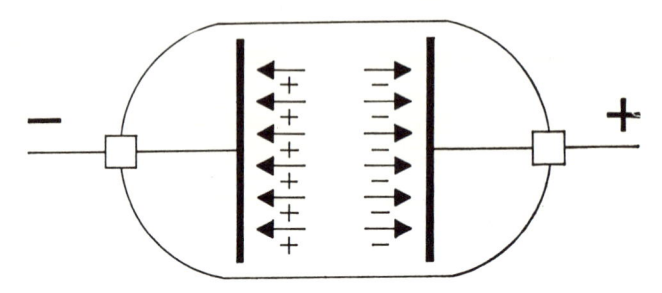

Fig. 14. Ionization chamber.

ions produced in the gas the greater is this reduction of the potential difference. The size of this reduction depends therefore on the number of particles which enter the chamber and on their ionizing power (Fig. 15).

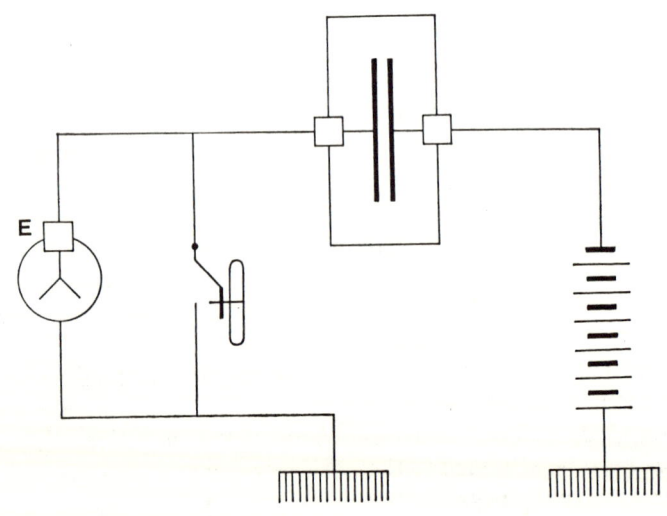

Fig. 15. How an ionization chamber is connected.

How Radioactive Particles are Detected 137

The ionization chamber is often connected to an amplifier (which may amplify about a million times) in order to produce an arrangement which is sufficiently sensitive to detect the passage of a single particle. The detection of the recording instrument which will be connected to the amplifier will be proportional to the ionization of the single particle in the gas, providing that the amplifier produces no distortion. This arrangement is called an *ionization chamber with a proportional amplifier*. It allows one not only to count the single particles which enter the chamber but also to measure the number of ion pairs which each individual particle produces in the gas of the chamber; and as a known energy is required for an ionizing particle to produce one pair of positive and negative ions one can also determine the energy lost by the particle in the gas of the chamber.

In recent years the technique of the ionization chamber with proportional amplifier has been very greatly improved and been developed towards what is known as the fast technique; that is the technique in which there is only an exceedingly short interval between the time in which the ionizing particle crosses the ionization chamber and the time at which the voltage pulse reaches its maximum value. The advantage of this technique is that it allows one, for example, to decide if two events which take place in two ionization chambers are contemporary or if, on the other hand, they are separated by a very short time interval, for example of some micro-seconds (millionths of a second). This fast technique demands the use of rapid chambers, of chambers that is in which the electrode connected to the amplifier collects only the electrons and not negative ions, since the latter will move slowly through the gas. The amplifiers should be fast, i.e. they should be built in order to transmit voltage pulses of very short lengths. The shortest times which are reached today with these systems are of the order of a tenth of a micro-second.

Finally, during the years 1947 and 1948 liquid and crystal ionization chambers were built. These solve the problem of having a large quantity of material in a small volume. A liquid ionization chamber is analogous to that of Fig. 14, except that it is filled with liquid argon. A crystal chamber on the other hand is essentially made up of an isolated crystal in which two opposite and parallel faces are silvered or gilded and connected to two soldered wires (Fig. 16). There is a potential difference of some hundreds of volts between these two plates. If an ionizing radiation passes through the crystal it excites the atoms and makes the crystal momentarily into a conductor.

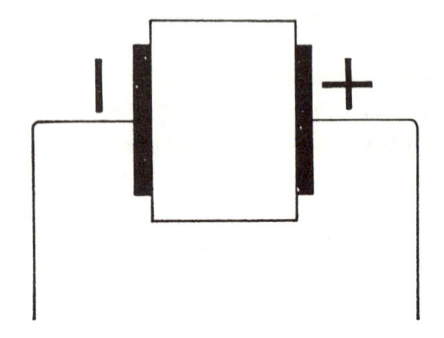

FIG. 16. Crystal ionization chamber.

The difference in potential between the two faces is reduced and the amount of the reduction is a measure of the intensity of the incident radiation. If these liquid and solid ionization chambers are connected to a suitable amplifier they can do more or less all that can be done by a gas ionization chamber. The choice of the instrument used depends naturally on the type of experiment to be carried out.

Another arrangement which is used to detect the passage of a single particle is the Geiger counter. This consists essentially of a metal tube T filled with gas, closed at the end by two insulators B (Fig. 17). The metal wire F is fixed to the insulators B and lies along the axis of the tube. Between the tube and the wire there is a high potential difference (1,000–2,000 volts)

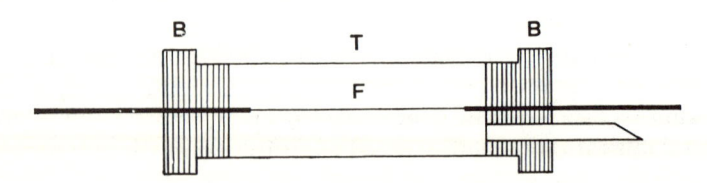

FIG. 17. A Geiger counter.

How Radioactive Particles are Detected

which is chosen so that in normal conditions it is just inadequate to produce a spark. If therefore a charged particle enters the counter the few ions which it produces in the gas rush towards the wire and because of the very high electric field acquire an energy which is large enough to ionize other atoms. The new ions in their turn do the same thing. The number of ions therefore increases very rapidly; in this way there is formed what is known as an *avalanche* of ions and so the discharge is produced and there passes momentarily through the counter a relatively high current. As a result the difference in potential between the wire and the tube falls to a value which is insufficient to produce further ionization; the discharge is interrupted and the counter returns to the initial condition; in this way it is ready to signal the passage of a successive particle. The discharge is amplified and then detected.

If the voltage applied to the counter is not very high and if the counter is connected to a proportional amplifier (which amplifies between a thousand and ten thousand times) the recording instrument gives a deflection which is proportional to the ionization produced by a single particle during its passage through the inside of the counter. An apparatus of this type is called a *proportional* counter; it supplies data entirely similar to those obtained with an ionization chamber with a proportional amplifier; however, whilst in the latter only the ionization produced by the incident particle comes into play, in the case of the proportional counter there is a multiplication of ions: the subsequent amplification can therefore be smaller.

Sometimes it is interesting to count how many charged particles pass through the counter without wanting to know the nature of the particles; one then uses a Geiger counter, that is a counter to which a relatively high voltage is applied and which is linked to an amplifier which amplifies only about ten times. In this instrument (which is also called a *relay counter*) the passage of a charged particle gives rise to a very large pulse (some tens of volts) whatever the particle may be (alpha, electron, proton . . .) When it is linked to an electromagnetic counter each impulse moves on one number.

Very often in place of a single counter two or more counters are used in *coincidence*: by this we mean an apparatus which, using suitable electronic circuits, registers a number on a suitable electromagnetic recorder when, and only when, a particle passes through all the counters. If for example three counters are in coincidence (Fig. 18) the numerical register connected to the output of the electronic circuit will be moved

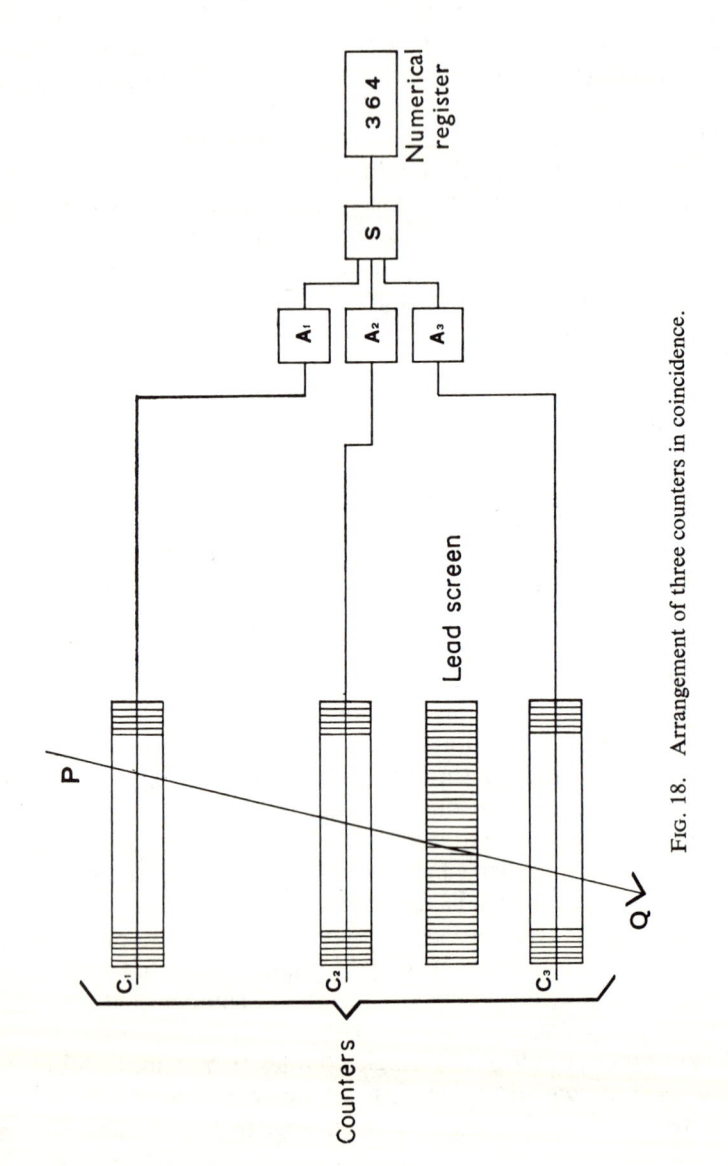

FIG. 18. Arrangement of three counters in coincidence.

on only by particles which have trajectories such that they cross the three counters in instants which follow each other by fractions of a millionth of a second. Particles which cross a single counter are not registered. So it is easy to understand that one can make a system of counters which register only those particles which come from a well-defined direction (Fig. 19).

The instruments which we have so far described (ionization chambers and counters) exploit the fact that a charged particle, when it passes through matter, *ionizes* the atoms which it encounters in its path. Other methods on the other hand make use of the fact that the particle *excites* the atoms along its path; that is, it causes the outermost electron to revolve in an orbit of higher energy; after some fractions of a second these electrons return to their initial orbit, emitting the energy that they have acquired in the form of light. There are two methods of detection based on this phenomenon: *scintillation counters* and *Cerenkov counters*, which amplify and register the very weak light sources produced by charged particles which cross matter due just to this excitation of atomic and molecular states.

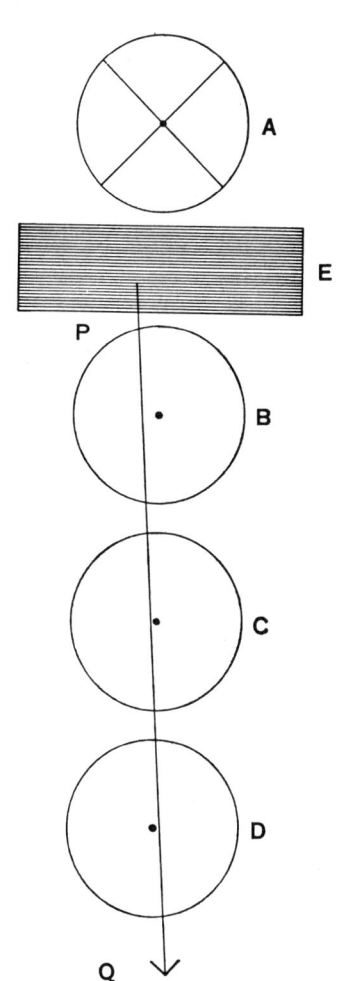

FIG. 19. Apparatus for the detection of a particle which originates in a lead screen.

Scintillation Counters

The scintillation method with which Rutherford and the Curies carried out their classic experiments appeared later to be superseded by the electrical method described above; in recent years, however, it has returned to the limelight but in a transformed guise.

The apparatus that was once used and which was called a *spinthariscope* (Fig. 20) consisted of a small glass screen covered with a thin layer, S, of zinc sulphide. On one side of the screen there was a microscope, M, whilst on the other side there was a support, R, on which a grain of radioactive substance was placed. The observer applying his eye to the microscope saw on the screen numerous scintillations, each of which was due to the collision of a single alpha particle against the atoms of the phosphorescent substance. In this way it is possible to count the

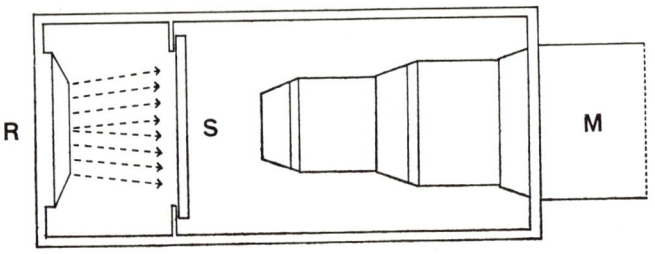

Fig. 20. A spinthariscope.

particles which fall in a given time on a given area. Protons also produce an effect which is analogous but less intense.

Everybody possesses a watch with a luminous dial; well, the luminosity of this dial is due just to this phenomenon of scintillation: mixed with a phosphorescent substance is a very small quantity of a radioactive material which emits alpha particles and causes the phosphorescent substance to give off light. By looking at the luminous dial with a magnifying glass one can distinguish the innumerable individual scintillations which, taken together, give the impression of constant luminosity.

This method of scintillations has been modified and improved so that it can also be used to detect beta rays and gamma rays. In *scintillation counters* one makes use of the fluorescence which crystals of sodium iodide or anthracene or stilbene exhibit as a result of a collision with particles or gamma rays; these substances fluoresce in ultra-violet light and, being transparent, can be used in thick layers. Special photo-electric cells called photomultipliers are used for observation and for counting. This method has the advantage of being exceedingly rapid so that it makes possible the investigation of phenomena which last only a period of time of the order of a thousandth of a millionth of a second.

Cerenkov Counters

In scintillation counters then there is amplified and registered the light emitted by the atoms which lie along the path of the particle and which re-settle themselves after having been excited by the collision. When however a charged particle crosses a transparent material with a velocity greater than the velocity of propagation of light in that material there occurs the *Cerenkov effect*. This takes its name from the Russian physicist who discovered it in 1934 and who in 1958 was awarded the Nobel Prize for this achievement.

If c is the velocity of light in vacuum and n is the refractive index of the material in which it propagates the velocity of light in this material is equal to the ratio c/n. When the velocity v of a particle in a material with refractive index n is greater than c/n the Cerenkov effect appears; this is an optical effect which is essentially analogous to the shock wave which is formed in air when a projectile or a jet plane moves with a velocity greater than the velocity of sound.

Let us imagine that the particle follows the track indicated in Fig. 21a) with a velocity greater than c/n; when it has reached E the surfaces of the light waves which (because of the phenomena of excitation) were emitted from the points A, B, C, D . . ., have reached respectively the spheres a, b, c, d . . .; these surfaces envelop the conical surface at which therefore the contributions of the various waves arrive in phase; in the direction perpendicular to this conical surface one has therefore effective light radiation. On the other hand in the case when the particle moves with a velocity less than c/n the situation is that represented in Fig. 21b; the particle will never cross the surface of the light waves which left the points on the trajectory through which it has passed; these elementary waves are superimposed in all possible phase relations in such a way that taken together the energy radiated by this means is negligible.

In a given material the Cerenkov effect is independent of the

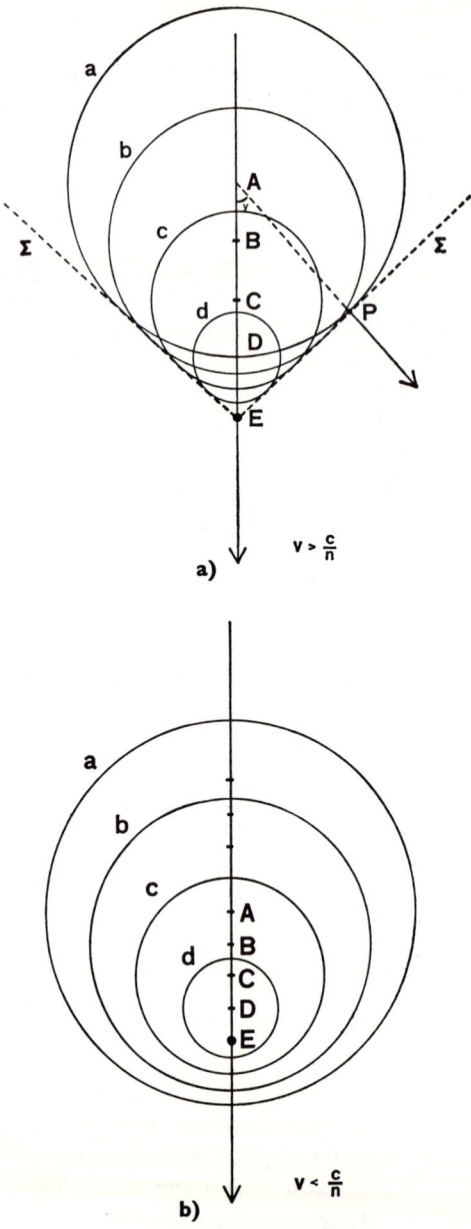

FIG. 21. The Cerenkov effect.

mass of the particle and depends only on its velocity and on its electric charge. And further, as contrasted to the scintillation phenomenon, it does not depend on the chemical composition or molecular structure of the material traversed but only on the value of its refractive index.

Making use of the characteristics of this phenomenon there have been built *Cerenkov counters* which are now in current use. They respond exceedingly rapidly and are particularly useful for determining the velocity of a particle and the direction in which it is travelling.

All of these instruments which are grouped together under the general name of *counters* are used to localize with great precision the instant of the passage of a sub-atomic particle; and according to the type and energy of the particles which are to be detected and according to the other information which one wants to obtain one uses one or other of the methods, all of which have been developed with a technique which from day to day becomes more and more refined.

I will now discuss briefly detectors which allow the observation of the tracks followed by a particle which crosses them (which make visible the position in space of every single ion produced by it) leaving however poorly defined the time *when* this passage took place. I will explain the individual phenomena on which are based the Wilson chamber, the photographic emulsion method and the bubble chamber.

TRACK DETECTORS

Wilson Chamber

The Wilson chamber, which has been called 'a window opened on to the atomic world' and 'the most brilliant and marvellous instrument in the history of science' was conceived in 1911 by the English physicist C. T. R. Wilson; in 1927 he was awarded the Nobel prize for the discovery of this elegant method of particle detection.

Before describing Wilson's apparatus we must put the question—what are clouds and how are they produced? The operation of the Wilson chamber (which is also called the cloud chamber) is based on the properties of clouds.

Clouds are made of a very large number of water droplets; at the centre of each droplet there is a grain of atmospheric dust. Atmospheric humidity cannot, in normal conditions, exceed a certain value at which the air becomes saturated with water vapour. This value depends on the temperature: the lower the temperature the lower it becomes. So when air which is saturated with vapour becomes cooler, a part of the vapour condenses and is converted into water. It may however happen that the air cools rather suddenly, for example when night falls. In this case the vapour does not have time to condense and the air remains supersaturated with vapour. Now supersaturated vapours are unstable and the presence of the slightest impurity produces condensation. So dust particles in the atmosphere act as centres of condensation; a droplet forms around each particle; these droplets are so small and light that they remain suspended in air. This explains the thick clouds which for long periods cover London and industrial cities in general. Multitudinous chimneys continuously discharge small impurities into the air which thicken the atmospheric dust; this dust provides condensation centres as soon as, as a result of particular climatic conditions, the air becomes supersaturated.

Let us return to the starting point of our discussion. Wilson realized that the ions produced by radioactive radiation served as ideal centres of condensation for supersaturated vapour. So when a sub-atomic particle passes through air containing supersaturated vapour, this vapour condenses about the trail of ions produced by the particle along its path. Thus the track of an ionizing particle which crosses this vapour becomes visible in the form of a thread, which appears bright, indeed brilliant, when illuminated by a powerful beam of light.

The Wilson chamber is used to produce supersaturated vapour which condenses around ions. It (Fig. 22) is essentially made of a large glass cylinder, inside which is a mixture of air and water vapour. An airtight piston moves inside the cylinder so that the volume available to the gas can be regulated. If the piston is pulled suddenly downwards, the mixture of air and water vapour expands and is therefore cooled. Water vapour which was originally saturated becomes now supersaturated and a passing particle leaves its trace in the form of a thread of cloud. It is for this reason that Wilson's apparatus is also called a cloud chamber; and it has been shrewdly observed that it is

How Radioactive Particles are Detected

not perhaps a fluke that it was invented in England, the country in which clouds are so common.

In order to photograph (with the cameras F_1 and F_2) the tracks of particles in the Wilson chamber it must be powerfully illuminated for a fraction of a second at the same time as the expansion. The threads of cloud which scatter light strongly, are recorded on the film and appear on prints as white on a black background. The tracks are photographed with a cine-

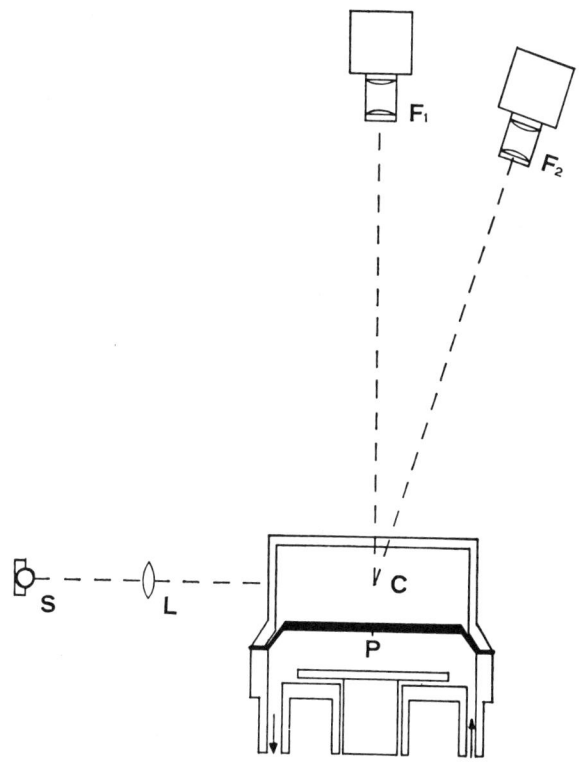

Fig. 22. The Wilson chamber.

camera; so, once the apparatus has been adjusted, one can take a large number of photographs. Furthermore, the chamber is equipped with an automatic apparatus which moves the piston at short and regular intervals of time and with another device, again automatic, which at each expansion removes the previously-formed ions, for example by attracting them on to charged plates.

A piston which rises and falls is relatively slow. In order to achieve greater speed one nowadays uses *membrane Wilson chambers*. The piston is replaced by a rubber membrane, and a sudden drop in pressure is produced behind the membrane; this then falls and produces an expansion of the gas in the chamber. In this way the parts which move mechanically can be made relatively light and the speed of the apparatus is increased. The chamber is filled with air or argon and the vapour is formed of a mixture of water and alcohol.

The Wilson chamber has, however, a serious disadvantage: the thread of cloud, which indicates the track of a particle, is produced only in the very short time (about one-tenth of a second) in which the gas is just at the point of saturation; but a minute or two is needed before the chamber is again ready to operate; for large Wilson chambers this time may reach several minutes. Thus most of the time is used only for preparation; and only once every few minutes is one able to get a rapid glance, which lasts a fraction of a second, at what is happening inside. For example, one hundred hours of work with a Wilson chamber which makes one expansion a minute, allows one to record particles which have crossed the chamber for a total time of only a few minutes. If, therefore, one wants to record a rare phenomenon, it is very unlikely that the event required will happen in the very short period of useful time; it would therefore be necessary to take an enormous number of photographs, hoping that a very lucky chance would allow one to observe the phenomenon under investigation.

In order to get over this disadvantage Blackett and Occhialini, in 1933, devised a very ingenious method in which the particle itself triggers the operation of the chamber. Schematically the method is this—a Wilson chamber is placed between two Geiger counters, C_1 and C_2, in coincidence (Figure 23); every coincidence pulse, which corresponds to a particle which has crossed the chamber, is used to operate a relay which sets in motion the apparatus for expansion and illumination. The arrangement may be complicated, containing more than two counters suitably placed around or inside the chamber. But the essential and elegant characteristic of the device is that it allows the

How Radioactive Particles are Detected

chamber to be operated only at the appropriate moment.

A barely trained observer can recognize at first sight which type of particle is responsible for a given thread of cloud in the chamber. The track of an alpha particle is thick, normally straight and rather short; that of an electron is thinner, longer and indicates a straight or irregular path according to the velocity. These differences are easily explained by the characteristics of the two particles: the alpha particle, which has a high ionizing power, makes more than ten thousand ions for every centimetre of its path in air and the cloud is formed around these ions; its track is therefore thick. On the other hand, the alpha particle is quickly braked when it crosses matter and so its track is short. Finally, since the alpha particle is heavy, it is undeflected by its collisions with atomic electrons and its track is normally straight. As we shall see, it is deflected only when it collides with a nucleus.

On the other hand, an electron of the same energy produces, as we already know, only about one hundred ions in a centimetre of path; its penetrating power is therefore much greater than that of an alpha particle and the track which it produces is thinner and longer; it is deflected frequently by collisions with nuclei and for this reason its track is irregular.

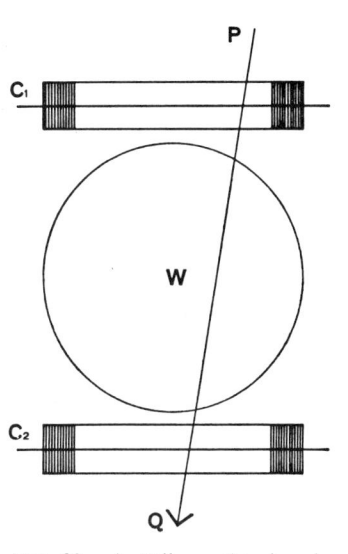

FIG. 23. A Wilson chamber in coincidence.

Sometimes one sees branching out from the track of a particle another lighter track; this second track also indicates the presence of ions. They are made by an electron with which the particle has collided. The ionization produced by this electron is called *secondary* ionization.

Looking with a Wilson chamber at a large number of tracks produced by alpha particles one encounters every so often one (about one in a hundred) in which one can observe a collision against a nucleus. An alpha particle colliding with a nucleus is deflected in its path and the struck nucleus is set in motion in a direction which forms a certain angle with that of the alpha particle track after the collision: one then sees in the Wilson chamber a track which, at a certain point, forks into two branches.

In general when two bodies collide together the direction of their motion after the collision makes an angle which is acute if the striking particle is heavier than that struck; it is obtuse in the opposite case and finally is a right angle if the two bodies have the same weight.

Of course this also happens for the collisions of an alpha particle with a nucleus. If the alpha particle strikes, for example, a nucleus of hydrogen which is four times lighter than the striking particle the two branches of the fork make an acute angle. Furthermore, from the thickness of the two branches one can distinguish which of them is due to the alpha particle: in fact, the alpha particle, having a larger mass, has also a higher ionizing power and therefore its track is thicker than the nucleus of hydrogen. If on the other hand an alpha particle collides against a nucleus of helium (that is, a nucleus of helium against a nucleus of helium) the two particles are equal: their tracks after the collision form a right angle. Finally, if an alpha particle collides with a nucleus of oxygen which is four times heavier the two tracks form an obtuse angle.

In special circumstances there are seen heavy tracks which start neither from the radioactive source nor from the walls of the Wilson chamber but from the middle of the chamber; such a track can be explained only by accepting that a *non-ionizing* particle has entered into the apparatus and by a collision has set in motion a nucleus which has then provoked the formation of the thread of cloud. We shall see later that the non-ionizing particle which is able to give a significant velocity to the nucleus is a *neutron*.

Electrons which are set in motion by photons, either by the photo-electric effect or by the Compton effect, may have an energy sufficiently high to produce ionization in the gas in which they are formed and in which they travel. One can understand therefore how an apparatus which indicates the presence of electrically charged particles may also indicate the passage of electromagnetic radiation of high frequency (gamma rays and X-rays). In fact in a Wilson chamber it is possible in certain cases to see weak tracks which do not come from the walls of the apparatus but which are born at a point inside the chamber; this indicates that at that point has occurred an interaction between a photon and an atom from which an electron has been liberated.

Sometimes a Wilson chamber is placed between the poles of a large magnet so as to deflect the trajectories of the charged particles which cross the chamber (see, for example, Plates 2 and 3); in these conditions not only is it possible to identify the sign of the charge of these particles (since the positive ones are deflected in a direction opposite to the direction in which the negative ones are deflected); it is also possible to determine their velocity (or more precisely their momentum); for the more slowly a particle moves the more it is deflected in a magnetic field.

The Photographic Emulsion method

A method of detecting particles has been suggested by Becquerel's initial experiment in which photographic plates recorded the radiation emitted by uranium salts. This technique of the *photographic emulsion* has been greatly developed in recent years and is very frequently used to reveal single ionizing particles and for the study of the phenomena to which they give rise.

When a particle (alpha particle, proton ...) crosses the emulsion of a photographic plate and passes close to a grain of silver bromide it produces in the usual way excitation and ionization processes which make it possible to develop the grain; thus, after the development and the fixing of the plate the track of the ionizing particle is distinguished by the alignment of the silver grains. In a Wilson chamber droplets are formed on the ions left by a particle and these droplets grow until they become visible; in a photographic emulsion around a certain number of negative ions (electrons) which are produced by the passage of ionizing particles there forms a small deposit of atoms of metallic silver which grow in the process of development so that they become visible in the microscope.

In order that this emulsion method should be efficient it is necessary to use specially prepared emulsions with exceedingly fine grains and to carry out the development and the fixing with a special technique in such a way that the gelatine is not deformed; it is moreover necessary to eliminate as far as possible the grains which would have been made sensitive for reasons independent of the passage of the particle. Again, there have been designed and produced emulsions which are sensitive to particular types of particles. The emulsions are examined using an ordinary microscope which is easily equipped for counting the grains along each track and for measuring lengths, angles, etc.

There are certain difficulties in the application of this extremely delicate method of detecting particles; first of all the scanning of an emulsion is very long and laborious; furthermore the appearance of the plates depends also on the time which elapses between the instant in which the plate is impressed and the instant of development because with the passage of time the tracks tend to disappear. But on the other hand the method has one extremely important advantage, which sometimes makes it indispensable: it requires no auxiliary apparatus, no electrical or magnetic fields nor moving mechanical parts. A small pack of photographic plates can be carried anywhere and by anyone; the development of the plates and the measurement of the tracks can then be done calmly in the laboratory. Furthermore with the emulsions which are prepared for the purpose by various firms it is possible to distinguish details as well as they can be observed with a Wilson chamber but with the advantage that it is possible to register in a permanent form rare events which occur over a long period of time.

This technique has led in recent years to the discovery of new particles and has revealed events of fundamental importance for the study of the atomic nucleus.

Bubble Chambers

Both of the two methods which render visible the tracks of ionizing particles—the Wilson chamber and the photographic plate—have advantages and disadvantages: a Wilson chamber is only sensitive for brief intervals and does not allow accurate measurement either of the density of the tracks or of their length; the photographic emulsion method demands a long, delicate and tedious exploration of the plate by means of a microscope; it does not give information on the time at which the event recorded in it took place and because of the high density of the emulsion, i.e. because of the too frequent atomic collisions, the track of the particle is so distorted that the effect of a magnetic field cannot be measured. And so the physicist Glaser asked himself in 1952 if it would not be possible to find a compromise between the two methods and so eliminate their respective defects and to combine their advantages; this implies using a substance which was dense enough to provide the advantages of the emulsion (frequent collisions by the ionizing particle and precise tracks) and yet not be too dense, so that as in the Wilson chamber the effects of a magnetic field could be recorded. It should in addition recover rapidly to re-establish operating conditions after every photograph. And Glaser had the idea of using as the material for his chamber a liquid and to exploit a property of liquids which had already been known for a long time. If all impurities have been eliminated from a liquid it can be heated to a temperature higher than its normal boiling point without boiling; but if in a liquid in this state a strange body is placed (for example, a piece of glass) it starts abruptly to boil violently. A superheated liquid exists in an unstable equilibrium such that a small cause can produce in it a large effect.

Glaser asked himself: if the small cause should be the transit of a sub-atomic particle would it produce along its path a visible effect? Well, theory showed and experiment confirmed that in a particular range of values of temperature an ionizing particle which crosses a superheated liquid produces along its path the formation of microscopic bubbles which give rise to the phenomenon of boiling (Fig. 24). This was the origin of the *bubble chamber*.

After the first experiments by Glaser others followed and the technique is being perfected; the tracks can be fine or thick according to the type of particle which produces them and the density of bubbles along a track is an indication of the ionizing power (that is, of the energy) of the particle which is better than the droplet density given by a Wilson chamber. But the enormous difference between a bubble chamber and a Wilson chamber consists in the number of rare events which can be observed; from this point of view a bubble chamber of fifteen centimetres

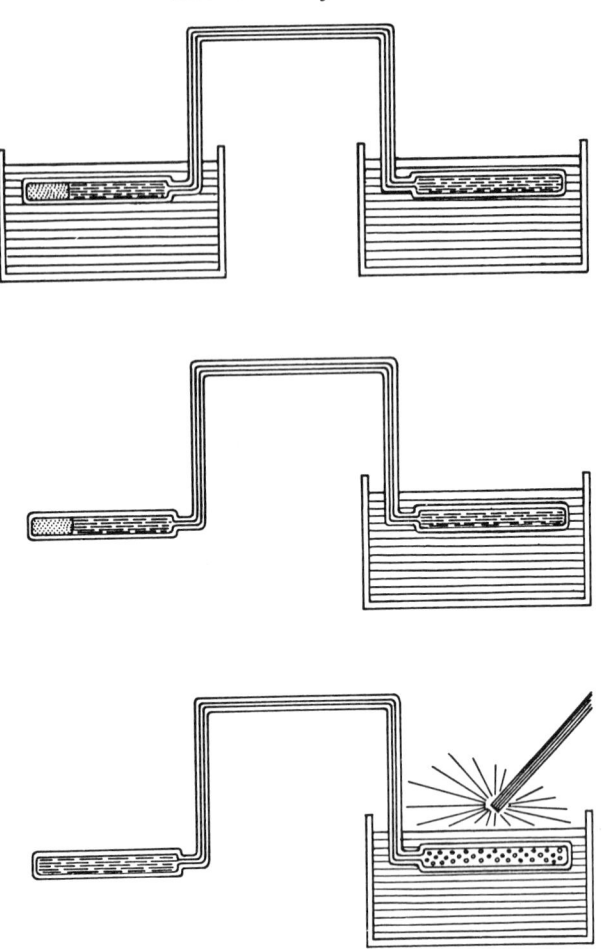

Fig. 24. The first experiment to show the possibility of making a bubble chamber; the bulb on the left, containing liquid ether, was warmed in a bath to 285°C; the thin tube between the two bulbs contained ether vapour. When the bath on the left was withdrawn the pressure in the tube fell from 21 to 1 atmosphere and the liquid ether in the right-hand bulb became superheated: it boiled as soon as a radioactive substance was brought near it.

is equivalent to a hypothetical Wilson chamber forty-two metres long! and finally the recycling time of a bubble chamber is very much shorter than that of a Wilson chamber.

Spark Chambers

An essential difference between the various types of counter on the one hand and the cloud chamber, bubble chamber and photographic emulsion on the other is that whereas the former give good resolution in time and relatively poor resolution in space for the latter the situation is reversed. The spark chamber is an instrument which provides good resolution in both space and time. First applied in experiments on high energy nuclear physics in 1960 it has now become one of the most important tools of the trade.

The spark chamber is a development of the *parallel plate counter*, an instrument similar in principle to the Geiger counter but employing parallel plates in place of the wire and cylinder electrodes. When an ionizing particle passes through a parallel plate counter the ensuing avalanche produces a spark. The current in this spark can be registered in an electronic circuit and thus the passage of the particle can be detected and 'counted'. But if the spark itself is photographed a segment of the track of the particle is revealed.

The spark chamber is formed of a series of parallel metal plates immersed in a noble gas, normally neon (Fig. 25). The distance between plates is usually a few millimetres and their

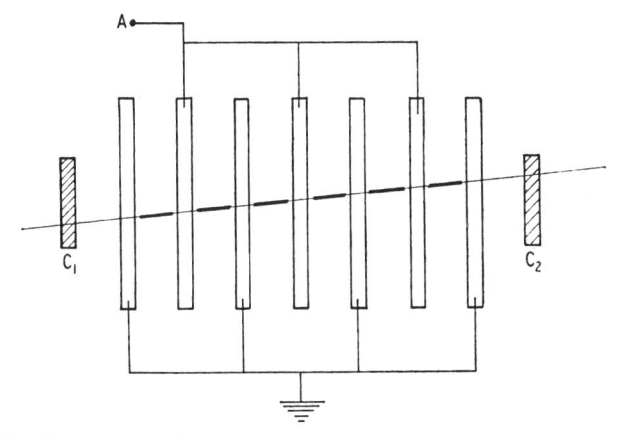

FIG. 25. Diagram showing the principle of a spark chamber. Alternate plates are connected electrically to A. When a particle passes through counters C_1 and C_2 a high-voltage pulse is applied to A. The ionization in the track of the particle leads to sparks being formed along the trajectory.

surface area may be any convenient size. Alternate plates are connected together electrically and, when opportune, a pulse of about ten thousand volts is applied to them from a circuit controlled by ancillary counters as in the counter controlled chamber. The trajectories of the particles passing through the chamber are revealed by a series of sparks (*see* plate 5).

A spark chamber photograph thus contains much more information than can be obtained from counters; at the same time, unlike the bubble chamber, the spark chamber can be triggered when an interesting event occurs—if necessary several times in one burst of particles from an accelerator. In practice the rate is limited by the speed of film movement in the camera; but it is not essential that the sparks should be photographed. The chamber can be scanned with a television camera and the spark positions recorded on magnetic tape for subsequent analysis by a computer. Alternatively the sound made by the sparks can be detected by suitably disposed microphones and located by sound ranging techniques.

In another development of the spark chamber the parallel metal plates are replaced by parallel arrays of closely spaced wires. It is then possible to identify the position of a spark by detecting the output pulses from individual wires; normally an array of vertical wires is followed by a horizontal array and with several of these pairs the track of an ionizing particle can be accurately recorded. It is of course also possible for the wires in one or two arrays as counters to trigger other arrays. 'Wire chambers' are a very recent development (they were first proposed in 1963) and have not yet been widely exploited; but it is probable that they will play an important part in the near future.

As we see each of the methods of detection has peculiar characteristics; one or the other is used according to the type of research which one wishes to execute.

These are the main methods of detection which the physicists of today have available; the marvel of these methods is that they make possible the study of the characteristics of the tracks of particles whose dimensions may be of the order of a tenth of a billionth of a centimetre.

CHAPTER SIX

THE ATOMIC NUCLEUS

NUCLEAR PHYSICS

In the study of atomic physics the structure of a nucleus is ignored. As it is so extremely small it is regarded as a material-point. Although this description is adequate to explain the greater part of atomic phenomena it is certainly incomplete. The most impressive demonstrations of this are the radioactive phenomena which consist in the spontaneous disintegration of the nuclei of some of the heavy elements. The existence of these radioactive phenomena leads us to assign to the nucleus a complex structure.

But what are the particles which make up the nucleus, this very small body whose radius is of the order of 10^{-13} centimetres, which contains almost all of the mass of the atom and which resides at the centre of the atom well guarded by the planetary electrons? And what are the forces which bind together the particles which make up the nucleus?

These are the problems which nuclear physics has tackled and has not yet resolved.

THE MASSES OF THE ATOMS

We have already noted the existence of isotopes, that is we already know that although only ninety-two elements can be distinguished chemically a much larger number (about 240) of different types of atoms exist in nature (if one takes account of differences of mass as well as of charge). There are some atoms which have the same number of electrons and are there-chemically indistinguishable but they are differentiated by different values of their mass and are therefore isotopes.

For example there is an element whose atom is made up of a

central nucleus, with a mass different from the mass of a proton, and of a single electron satellite. This is an isotope of hydrogen and has been called deuterium; its nucleus is called a deuteron. We shall see later the particular importance of this nucleus.

In order to measure the masses of atoms we use instruments which have the generic name of mass spectrographs or mass spectrometers. Many of these are based on a method which had been used by J. J. Thomson to measure the ratio between the electric charge and the mass of the electron. It consists in observing the deflection of a beam of positive ions of the element under investigation by an appropriately chosen combination of an electric field and a magnetic field. The original method of Thomson was made very much more precise by Aston whose mass spectrograph has contributed greatly in extending our knowledge of atoms. I will how describe the principle on which it is based.

Let us suppose we pass a beam of ionized atoms of a certain element through an electric field or a magnetic field and that all the atoms in the beam have the same velocity. As a beam of electrically charged particles is equivalent to an electric current it will be deflected in its trajectory and the lighter the atom the greater will be the deflection.

If the ions which make up the beam were all to have the same weight then they would all be deflected equally; but let us suppose for example that the element is made up of a mixture of two isotopes, that is of two types of atoms which have different weights although they both have the same number of external electrons. Now an atom when it loses an electron suffers a change in charge but hardly any change in mass. The ions which make up the beam will therefore also be of two different types with different weights. The electric field will therefore produce a separation of the two types of ion; the lighter ions will be deflected by a greater amount. Thus it we place a photographic plate in a suitable position we will find on it two distinct traces due to the two beams of ions, one corresponding to the heavier ion and one to the lighter ion. One can derive the weight of the two isotopes of the element being investigated from the position of the two marks on the photographic plate. This method can naturally be applied whatever the number of isotopes which make up the element being investigated.

The Atomic Nucleus

This is the principle on which the mass spectrograph is based. The ions are passed first through an electric field and then through a magnetic field (which is produced by an electromagnet). The direction and the values of these fields are such that all the ions which have the same ratio of electric charge to mass are focused in a point whatever the value of their velocity.

Using this instrument Aston confirmed and extended Thomson's results. Nearly all of the elements are a mixture of isotopes. The atoms of each of these have nuclei whose mass is an integral multiple of the mass of the proton, that is of the hydrogen nucleus: this integral number is called the mass number. For example, iron is a mixture of three isotopes which have mass numbers 54, 56 and 57: this means that the atomic nucleus of one of these isotopes has a mass which is fifty-four times greater than the mass of a proton, the nucleus of another isotope fifty-six times and finally the nucleus of the third has a mass equal to fifty-seven times the mass of a proton.

The nuclei of two isotopes (for example the nuclei of the two isotopes of iron with mass numbers 56 and 57) differ one from the other just as they differ from the nucleus of the previous element (manganese) which has a mass number of 54. From the nuclear point of view the fact that the two isotopes 56 and 57 have the same chemical properties, that is that they are both iron, is but a secondary effect which depends on the number of electrons which surround the nucleus—a number which for the two isotopes of iron is the same. Somebody who met in the street a group of soldiers might say looking at their uniforms, 'here is a group', for example, 'of parachutists'; but the common uniform clothes different individuals—as different from each other as one of them is different from the person who observes them. Thus for a chemist, the equivalence in chemical properties makes the isotopes of the same element indistinguishable; he calls them all by the same name, but from the point of view of the nuclear physicist they are as different as are elements with different names; the cloud of external electrons is only a uniform which conceals the real individuality of the atomic nucleus.

In order to identify a particular isotope of an element one uses the same letter which serves in Mendeleev's table to indicate the elements; and one adds at its side two numbers: one on top, and to the right, to indicate the mass number (indicate, that is, the ratio of the mass of the nucleus to the mass of the proton); the other below, and to the left, indicates the atomic number (that is the number of planetary electrons in the atom which is naturally equal to the value to the positive electric charge of the nucleus.) For example, the symbol $_7N^{14}$ identifies the nucleus of the isotope of nitrogen which has the atomic weight 14; the atomic number of nitrogen is 7.

I have said that it was discovered from Aston's measurements that the masses of the atomic nuclei of isotopes are integral multiples of the mass of the protons. Now this is not exactly true, for when later Aston increased the resolving power and the precision of the apparatus he realized that this integral number rule was not in fact precise. The masses of the atoms are almost, but not exactly, represented by integral numbers.

For example, whilst Aston's first measurements gave for the masses of the two isotopes of chlorine 35·0 and 37·0 the later, more accurate measurements showed that the values of these masses are in reality equal to 34·983 and 36·980, that is they are very slightly less than integral numbers.

One defines then as the mass number of an isotope the integral number which is closest to its atomic weight: thus 35 is the mass number of the chlorine which has atomic weight 34·983 and 37 is the mass number of the chlorine with atomic weight 36·980.

Oxygen is made up of three isotopes of mass 16, 17 and 18 which are indicated using the convention which we have just made as O^{16}, O^{17} and O^{18}. They make up respectively 99·76%, 0·04% and 0·20% of the oxygen which is found in nature; as a result of this discovery it has been necessary to revise the unit of atomic weight.

Two scales are needed: the chemical scale of atomic weight, in which (as was done before the discovery of isotopes) one sets equal to 16 the mass of the element oxygen which is found in nature; and the physical scale of atomic weight (which is used in nuclear physics) is formed by setting equal to 16 the mass of the lightest of the three isotopes of oxygen which exist in nature; that is one puts $O^{16} = 16$. On this physical scale one has

Atomic Weight of the electron	Atomic Weight of the proton	Atomic Weight of Hydrogen
0·0005	0·0076	0·0081

THE DISCOVERY OF NUCLEAR TRANSFORMATIONS

The discovery of isotopes reawakened interest in the ideas of Prout who proposed in 1816 that all elements were made up of assemblies of the simplest element, that is of hydrogen. Physicists, struck by the fact that all atomic masses have values which are very nearly integral numbers, asked themselves: might not all of the different nuclei be made up of one or more common constituents?

There was only one way to answer this question: to try to break up an atomic nucleus. Nature has provided us with examples of the break-up of atomic nuclei in the phenomenon of natural radioactivity; here we can witness the spontaneous change of one element into another, that is the spontaneous disintegration of an atomic nucleus.

But we can only stand by and watch this phenomenon; we possess no apparatus with which we can accelerate it or slow it down. All the means which we have available (chemical reactions, high temperatures, electrical discharges, etc.) act only on the planetary electrons and leave untouched the atomic nucleus.

The alchemist's attempt to make gold by heating various metals (that is to change another metal into gold) is, translated into the language of modern physics, an attempt to modify the nucleus by using agents which act only on the electrons. The likelihood of success for the alchemist (in trying to convert other metals into gold) was therefore no greater than would be that of a mining engineer who tried to arrive at the centre of the earth using apparatus which was designed for digging a well.

In order to succeed in reaching the nucleus and to break it up into its constituent parts it was necessary to find a way of breaking through the crowds of satellite electrons. It was again the phenomenon of the natural disintegration of the atomic nuclei of radioactive elements which suggested to physicists that nuclei might be disintegrated by bombarding them with high energy particles. They needed therefore projectiles which had a high enough energy to be undeviated by the electric charge of the planetary electrons and of the nucleus itself and which would retain enough energy to break up the nucleus in a collision.

Physicists express the energy of radioactive particles in a unit which was introduced for practical reasons. They defined the 'electron volt' as the energy of an electron which has been accelerated through a difference of potential of one volt. It is given the symbol ev and is equal to $1 \cdot 6 \times 10^{-12}$ ergs; it is a unit which is now used throughout nuclear physics. For example, a particle is said to have an energy of a million electron volts (one Mev) when it has the same energy as an electron which has been accelerated through a difference of potential of a million volts.

As particles which are emitted in radioactive transformations have energies which are of the order of some millions of electron volts (i.e. of some Mev) it was thought necessary to use projectiles of about the same energy in order to break up an atomic nucleus. What could be more natural then than to use alpha particles which had been emitted by radioactive substances as projectiles with which to bombard other nuclei and to alter their structure. For many years, as we shall see, the only source of particles with sufficient energy were just radioactive disintegrations.

Rutherford was the first (in 1919) to bombard a gas, nitrogen, with alpha particles hoping that these projectiles would reach the nuclei and break them up. There was, of course, no method of focusing and therefore many of the projectiles were lost; however, some of them hit the target.

Rutherford's apparatus consisted of a container which was filled with a gas under examination. In one of the walls of the container there was a hole which was covered on the inside by a thin sheet of silver. A small glass screen covered with zinc sulphide was put outside the apparatus at the exit of the hole. The alpha particle source which was again of radium C was put on an appropriate support inside the container. Alpha particles could not reach the screen because they would be absorbed in the thin layer of silver.

When the container was filled with nitrogen scintillations were seen on the screen. Thus the nitrogen had given up particles which had enough energy to pass through the layer of silver and produce scintillations by collisions on the screen. In order to find out what these particles were Rutherford examined them in a magnetic field and concluded that they were protons, that is hydrogen nuclei. Later Rutherford and Chadwick were to show that various other light elements could be transformed in a similar way by emission of protons.

When, later, we consider artificial radioactivity we shall study in greater detail the results of this experiment. What concerns us now is the fact that it shows that nitrogen and also other elements emit protons when they are bombarded by alpha particles. From this it is logical to conclude that the proton is one of the constituent particles of the atomic nucleus.

Thus we now have two so-called elementary particles, that is two entities which give rise to the whole physical world: protons and electrons.

One might at that time have thought—and indeed for a short time it was thought—that all matter is made up of protons and of electrons; that is, that every atom is made up of a nucleus (positively charged) around which rotates a certain number of electrons; and that the nucleus, in its turn, is made up of protons and electrons. For example, the helium atom would be made up of two planetary electrons surrounding a nucleus which, in its turn, would be made up of four protons and two electrons. The four protons would account for the mass value of 4 and the two electrons, with their negative charge, would neutralize the positive charge of two of the protons and would thus account for the charge value of 2 for the helium nucleus.

But the rapid development of the theory very soon showed that one could not think that matter is made up only of protons and of electrons. In fact, Dirac in 1927 formulated the relativistic wave equation of the electron and showed theoretically that it is impossible to contain an electron in a region of space as small as that of a nucleus. Moreover, the hypothesis that the nucleus is made of protons and of electrons was shown to conflict with certain experimental results.

This difficulty was overcome in 1932 when, as we shall see immediately, the Englishman, Chadwick, discovered a new particle, the neutron, which is electrically neutral and which has the same mass as the proton.

DISCOVERY OF NEUTRONS

The problem of the artificial disintegration of nuclei immediately became of the greatest importance, and there were many physicists who devoted themselves to this subject as soon as Rutherford had published the results of his experiments. Amongst the many experiments that were made that of the German physicists Boethe and Becker in 1930 had a particular importance. Its results provoked a fundamental change in ideas on the structure of the atomic nucleus.

Bothe and Becker wanted to discover whether the artificial disintegration of nuclei was like the natural disintegration of radioactive nuclei accompanied by the emission of gamma rays. They therefore bombarded with alpha particles a certain amount of beryllium. The beryllium and the radioactive source were surrounded by a layer of two millimetres of brass which was enough to stop the alpha particles and any protons which might be emitted from the beryllium; gamma rays, on the other hand, could pass through the brass and be detected by a counter.

This experiment showed that beryllium when bombarded by alpha particles emitted a very penetrating radiation (it could cross layers of lead more than 10 centimetres thick). After an investigation of this radiation Bothe and Becker concluded that it was of the same general type as gamma rays but of a very much higher energy.

In 1932 Irene Curie (the daughter of Marie Curie) and her husband F. Joliot repeated the experiment, using an ionization chamber but they varied the type of screen which was used to stop the particles with low penetrating power. They discovered that when this screen was made of a substance containing hydrogen (water or paraffin) the current in the ionization chamber increased by an enormous amount. It was logical to think that the gamma rays in crossing the screen succeeded not only in ionizing some of the atoms of hydrogen but also in setting in motion the protons that were produced in this way. This effect could not be observed with metal screens because metallic atoms are very much heavier than protons and it is therefore much more difficult to put them in motion.

At first sight this explanation appeared satisfactory. When, however, the results were examined more closely and detailed calculations were made it was realized that electromagnetic radiation would have to have an impossibly large energy in order to be able to set protons in motion and to give them the velocity which was discovered by experiment. The Joliot-Curies therefore could see no plausible explanation of their results. They published them and explained the difficulties with which they were confronted.

The credit of overcoming this difficulty was left to the Englishman, J. Chadwick. He immediately repeated these experiments in the Cavendish Laboratory at Cambridge. He obtained results which were even more difficult to reconcile with the hypothesis that the penetrating radiation emitted by beryllium was of an electromagnetic nature; the radiation was in fact able to give high velocity to the nuclei of other light elements.

From the data obtained in these experiments Chadwick deduced that the protons and other nuclei are not set moving by electromagnetic radiation but by a new radiation which was made up of material particles which had no electric charge and which had a mass almost equal to that of the protons. He did this less than fifteen days after the announcement of the results of the Joliot-Curie experiments.

As these particles are electrically neutral they are called neutrons.[1] It was discovered that on the physical scale of atomic weights the atomic weight of a neutron is equal to 1·0090.

Beryllium when bombarded by alpha particles does not emit protons but neutrons and gamma rays. It was only the latter which were detected by the counter in the experiment of Boethe and Becker.

This then is the intriguing story of the discovery of the neutron. It was first unwittingly observed by one group of physicists whilst it was later recognized by other experimenters.

Here, then, we have another particle, hitherto unknown, which could be emitted by an atomic nucleus when bombarded by projectiles. We can therefore say for the time being that

---

[1] Twelve years earlier when it was thought that atomic nuclei were made up of protons and of electrons it appeared reasonable to think that the simplest nucleus would be made up of a single proton and a single electron, much more closely linked together than they are in atoms of hydrogen. Such a nucleus, whose existence, as we have seen, was later shown to be theoretically impossible, had been given by Rutherford the name of neutron.

protons and neutrons are two types of particle which must enter into the construction of the atomic nucleus.

The principle characteristic of neutrons is their ability to cross very large thicknesses of matter. This is due to their lack of electric charge.

For when a positively charged particle, for example an alpha particle, crosses matter it exerts on an electron even at a distance an attractive force of an electrical type. When this particle passes near a nucleus it exerts also on it an electrical force which, however, in this case will be repulsive because both the atomic nucleus and the alpha particle are positively charged. One might almost say that the alpha particle collides, even at a distance, with electrons and with a much smaller number of nuclei. The slowing down of electrically charged particles which occurs when they cross matter is due just to these electrical collisions.

Neutrons, on the other hand, as they do not carry charge do not feel in any way the effect of electrical forces and when they cross matter they are not slowed down even by an effective collision with an electron, which has a very small mass. The only thing which slows them down is an effective collision against a nucleus. As a result, the penetrating power of neutrons is about ten thousand times greater than that of alpha particles.

We shall see later that neutrons, being neutral, are extremely efficient as projectiles for producing artificial disintegrations of nuclei.

Neutrons do not exist in a free state because whenever they are in the presence of matter they react with nuclei and give rise to various types of transformation; if on the other hand they are in vacuum they decay according to a process which we will discuss later.

PROTONS AND NEUTRONS IN THE NUCLEUS

When Chadwick discovered the neutron 'physics had a luminous moment during which it seemed that nature had assumed a magnificent simplicity' (Marshak). It appeared that the physical universe could be explained in terms of only three elementary particles: electrons, protons and neutrons combined in different ways in ninety-two types of atom.

The Atomic Nucleus

An atom would be made up of a dense nucleus consisting of protons and neutrons and of a crowd of electrons revolving around the nucleus as planets revolve around the sun. The number and the arrangement of these electrons explain the chemical properties of the atom whilst the nuclear particles—protons and neutrons—account for its other properties such as its weight, its nuclear charge, its isotopes, etc. Electrons, protons and neutrons appeared to be just those irreducible material entities which had been sought by scholars from Democritus on.

But, as we shall see, this situation has become more and more complicated; as time went on new particles were discovered: positrons, neutrinos, negative protons, the large family of mesons . . .; there are more than thirty particles: there are very much too many. As has been said physicists are, in front of these growing complications, in the same situation as that of astronomers before Copernicus who, in order to retain the earth as the centre of the universe, were obliged to complicate and entangle more and more the movements of the planets as the measurements became more and more accurate. By giving up the dogma of the immobility of the earth (inherited across the centuries from the Greeks) astronomers were able to introduce simplicity and harmony where previously chaos was dominant. Will the current situation in the field of nuclear physics be cleared up in the same way? Today one can only express this hope and this wish.

Despite, however, the existing complicated situation physicists have now reason to think that the atomic nucleus is effectively made up only of protons, of neutrons and of nothing else. That is, in other words, that protons and neutrons belong to a class which is different from that of other particles. They are the bricks from which the nucleus is made (and for this reason they are given the generic name of nucleons), whilst other particles have a different role in the physical world; I will return to this later.

Thus the different nuclei are made up of protons and neutrons, that is of two types of particles of almost equal mass, one positively charged, the other neutral.

Let us remember that to a good approximation a proton has mass one and charge one, and a neutron has mass one and charge nought. Thus all of the charge of a nucleus is necessarily due only to the protons. If the nucleus has, for example, charge four this means that it contains four protons. The mass of a nucleus, on the other hand, is due both to protons and to neutrons. Thus, if, for example, a nucleus has mass eight, this means that the combined number of neutrons and protons is eight.

Let us now see how any nucleus is made up, for example a nucleus of aluminium. Aluminium has atomic number 13 and atomic weight 27; as its atomic number coincides with the nuclear charge the nucleus of aluminium has charge 13 and therefore should contain 13 protons. On the other hand, the combined number of protons and neutrons is equal to the atomic weight 27 and thus the number of neutrons is equal to $27 - 13 = 14$. We conclude that the nucleus of aluminium is made up of 13 protons and 14 neutrons.

In the same way one can go on to discover the number of protons and of neutrons of all elements. The atomic number Z gives the number of protons, the difference, $A - Z$, between the mass number, A, and the number, Z, of protons tells us the number of neutrons.

In particular the hydrogen nucleus which has the mass number 1 and the atomic number 1 is made up of a single proton without a neutron. There is also a nucleus which was discovered in 1932 by the American Urey which still has atomic number 1, like hydrogen, of which it is, therefore, an isotope; however, its atomic weight (that is its mass number) is 2, it is therefore made up of a proton and of a neutron. The name of *Deuteron* is given to this very simple assembly.

As I have already said, the atom of that isotope of hydrogen whose nucleus is a deuteron (instead of a proton) is called deuterium or heavy hydrogen and is given the symbol D. When one speaks of heavy water one refers to a molecule of water in which the normal hydrogen is replaced by deuterium, that is a molecule of type D_2O. Heavy water and deuterium are very widely used today for research in nuclear physics, in physical chemistry and in biology.

Deuterium is a very rare isotope of hydrogen. In 6,000 atoms of hydrogen one finds only one of deuterium, but there exists another isotope of hydrogen which is even more rare than deuterium. It is tritium with atomic weight equal to 3. Its nucleus has the name of triton (a name which has no reference to mythology!). As we shall see the deuteron and the triton are the main fuels which one hopes to be able to use in thermonuclear reactors (Fig. 26).

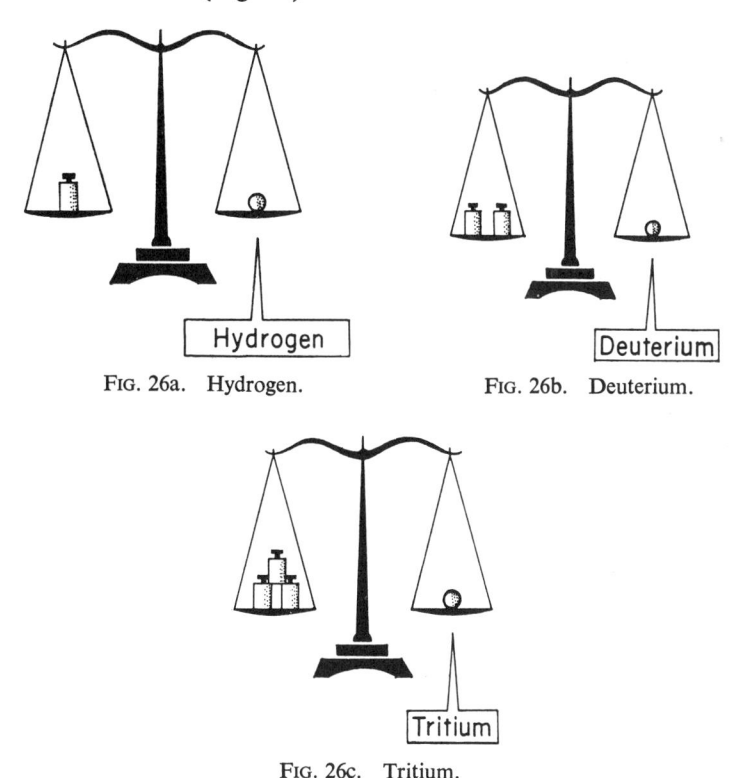

FIG. 26a. Hydrogen.

FIG. 26b. Deuterium.

FIG. 26c. Tritium.

NEUTRON PROTON DIAGRAM

Just as Mendeleev's system gives one a synthetic picture of the elements which exist in nature arranged according to their chemical properties so one might think of making a similar table for the nuclei of all the atoms which exist in nature; but whilst in the case of Mendeleev's table there is only one quality which varies from element to element, that is the atomic number (equal to the number of planetary electrons of the atom, that is equal to the value of the charge of the nucleus) in the case of nuclei one has to take account of two factors, of the charge and of the mass of the nucleus, that is of the number of protons and of the number of neutrons which make it up.

A graphical representation is very much more vivid than a simple table. Let us draw a diagram on which we put on the abscissa the number, Z, of protons and on the ordinate the number, $A - Z$, of neutrons (Fig. 27). The different isotopes of the same element have, just because they are isotopes, the same atomic number and thus the same number of protons. They will therefore be represented in the diagram by points on the same vertical line.

As one looks at the diagram one immediately notes a very important fact. All the points representing the nuclei of different elements are grouped along a narrow strip which defines the region of nuclear stability. Furthermore, the lighter nuclei are made up of the same number of protons and neutrons; but as one moves up to nuclei of higher atomic number one finds that the number of neutrons in the nucleus is always greater than the number of protons. The heavier the nucleus the more marked is this effect. The nucleus of the isotope 238 of uranium (which is the heaviest of the elements) is made up of 92 protons and $238 - 92 = 146$ neutrons.

This excess of neutrons over protons in the natural heavy nuclei is due to the fact that the protons, being positively charged, exert repulsive forces on each other which inhibits the accumulation of protons in heavy nuclei.

THE MASS DEFECT OF A NUCLEUS

Precise measurements of the atomic weights of the various isotopes have shown that the mass of the nucleus which, as we know, has a value very close to an integral number, the mass number (i.e. the number of protons plus the number of neutrons) is not equal to the sum of the masses of the nucleons (protons and neutrons) which make it up. It is smaller by some ten parts in a thousand. This difference between the value of the sum of the masses of the constituent particles of the nucleus and of the effective value of the nucleus itself is called the *mass defect*. Why does this mass defect exist? Where has this small amount of missing mass gone?

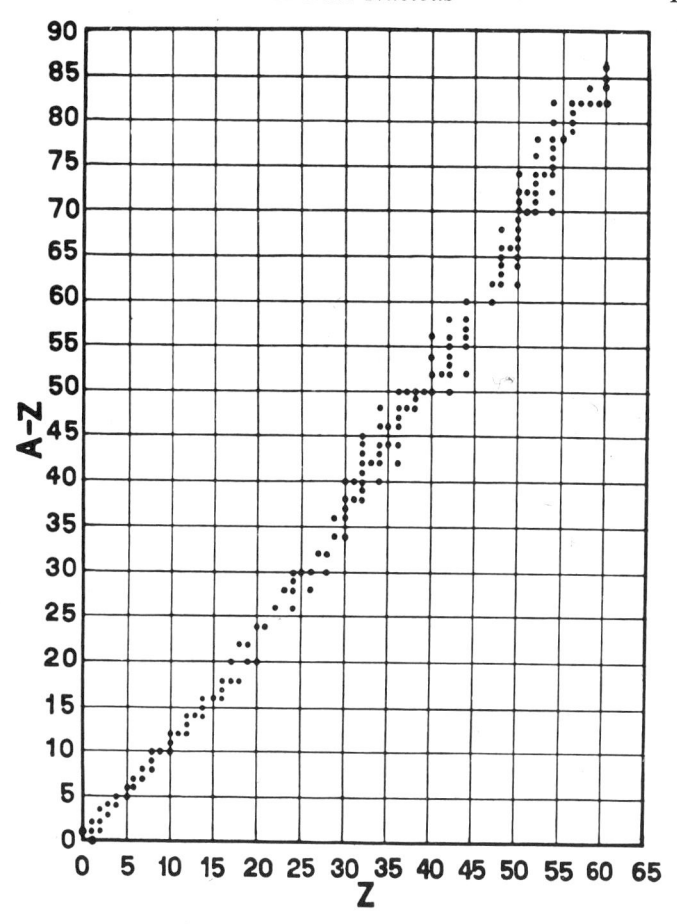

FIG. 27. Proton-neutron diagram.

The reply to this question is supplied by Einstein's theory of relativity. One of the most important consequences of this theory is that mass and energy are equivalent, in the sense that mass can be changed into energy and vice versa. In detail, every time that in a physical process one records the disappearance of a mass, M, there appears (in one or other form) an energy, E, which is related to M by Einstein's famous equation

$$E - Mc^2$$

where c is the velocity of light in vacuum (about 300,000 kilometres a second or 3×10^{10} centimetres a second).[1] If M is expressed in grammes and c in centimetres a second E will be given in ergs.

Thus the mass of one gramme represents an energy of $(3 \times 10^{10})^2$ or 9×10^{20} erg; thus if the whole of the mass of one gramme is transformed into energy it would be equal to 900 million billion ergs. This is an exceedingly large amount of energy. In order to obtain it by burning coal one would need 3,000 tons.

I said that the unit of atomic mass (one-sixteenth of the mass of an atom of the isotope 16 of oxygen) is equal to the mass of $1 \cdot 660 \times 10^{-24}$ grammes. Thus every unit of atomic mass is equal to an energy

$$E = 1 \cdot 660 \times 10^{-24} \times (2 \cdot 9978 \times 10^{10})^2 \text{ erg} = 932 \text{ Mev},$$

when we remember that in nuclear physics the energy is normally expressed not in ergs but electron volts and that

1 electron volt = 1 ev = $1 \cdot 6 \times 10^{-12}$ erg

1 mega electron volt = 1 Mev = $10^6 \times$ ev = $1 \cdot 6 \times 10^{-6}$ ergs

As a result every atomic nucleus which has a mass M (expressed of course in units of atomic mass) is equal to an energy of $932 \times M$ mega electron volts.

In order to illustrate further the relation between mass and energy I will add that when a particle (for example an electron of mass, M) has a total energy, W, this energy is necessarily always greater than mc^2 which represents what is called its *rest mass*. The difference

$$\text{W} - Mc^2$$

represents on the other hand its *kinetic energy*.

THE BINDING ENERGY

Einstein's relation then supplies the answer to the question which we asked ourselves: what happened to the missing mass in an atomic nucleus?

We know that protons and neutrons are in stable nuclei

---

[1] More precisely the velocity of light in vacuum is equal to $2 \cdot 9978 \times 10^{10}$ centimetres per second.

tightly bound together. Well then, the mass which has apparently disappeared when they are strongly linked to form the nucleus is turned into energy and it is just this energy which holds them bound together; it is this which is called the binding energy of that nucleus.

Let us consider for example a helium atom in which the nucleus is made up of two protons and two neutrons. It has been found that its atomic weight is equal to 4·0040; but remembering that the atomic weight of hydrogen is 1·0081 and that of the neutron is 1·0090 the combined weight of the particles which make up the atom of helium (two atoms of hydrogen and two neutrons) is 4·0342. Thus the mass defect of helium is equal to $4·0342 - 4·0040 = 0·0302$ units of atomic weight which corresponds to a binding energy equal to $0·0302 \times 932 = 28$ Mev.

We shall see later how it has been possible to liberate and to use this energy which inside atomic nuclei holds tightly bound together its constituent parts: protons and neutrons.

THE POSITRON

After the discovery of the neutron had brought up to three the number of particles which make up atoms it appeared, as I have said, that the problem of the constitution of atoms had been resolved with beautiful simplicity; but this period of happy illusion was in practice exceedingly short. In less than one year after the discovery of the neutron there was discovered a fourth elementary particle.

In the course of 1933 the American physicist, C. D. Anderson, was carrying out at the California Institute of Technology certain experiments on cosmic radiation (to which I will refer later). The experiments were made to measure the kinetic energy of the particles in cosmic radiation. To do this he observed their tracks in a Wilson chamber which was set up in a magnetic field. The effect of this field, as we know, is to curve the trajectory of a charged particle and the lower the energy of the particle the more curved will this trajectory be. One can understand therefore how Anderson, from the curvature of the trajectories of the cosmic ray particles, could arrive at their energy. On the other hand, a magnetic field deflects positively charged particles in one direction and negatively charged particles in the opposite direction.

Anderson realized that not all the trajectories were curved in the direction corresponding to negative charge (electrons) but that some were curved in the opposite direction and thus indicated the passage of a positively charged particle (Fig. 28). He thought at first that the positively charged particles were protons of exceedingly high energy but as he carried on the experiment he discovered that some positively charged particles had a mass which was less than that of a proton and equal to that of the electrons. Here there was a new type of particle which was called the *positive electron*, or *positron*.

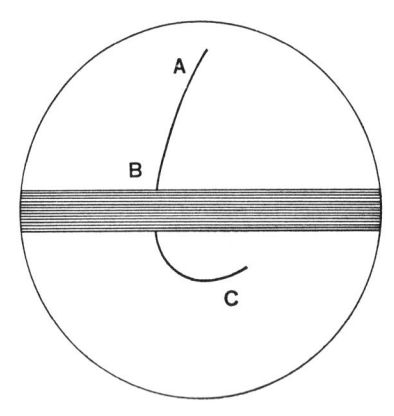

FIG. 28. Sketch of the photograph which enabled Anderson to discover the positron.

Almost at the same time, still in 1933, the Englishman, P. M. S. Blackett, and the Italian, G. Occhialini, carried out brilliant experiments in this field. They studied the cosmic radiation and were able to photograph exceedingly complicated phenomena in which one saw, emerging from the same point some tens of tracks of very high velocity particles (showers). If this phenomenon is observed in a magnetic field one can see that some of these tracks are due to normal negative electrons whilst others are deflected in a direction such that they must be attributed to electrons which have instead of a negative electric charge a positive charge: that is, they are positrons.

CREATION AND ANNIHILATION OF ELECTRON-POSITRON PAIRS

After the discovery of the positron its existence was recognized not only in phenomena due to cosmic rays but also in other phenomena which could be created in a laboratory. In particular, it was discovered that when matter was irradiated with high energy photons (gamma rays) pairs made up of an electron and a positron were produced. This phenomenon occurs only when the energy of the incident photon is greater than about one Mev and consists in the disappearance of a photon and the simultaneous creation of a pair of electrons, one positive and the other negative.

This phenomenon takes place in a sufficiently strong field of force, such as the electric field produced by a nucleus in its immediate surroundings. It makes up one of the most elegant proofs of the equivalence between mass and energy which had been found by Einstein.

Following a proposal by Marie Curie these electrons and positrons which are formed by the expenditure of energy are called *materialization electrons*.

A process which is opposite to this is the *dematerialization* of particles. This effect consists in the annihilation in the neighbourhood of a nucleus of a positron-electron pair with the production of two or three gamma quanta.

The experimental discovery of the positron is a triumph for theoretical physics. The existence of the positron had been predicted two years earlier by P. A. M. Dirac as a result of his relativistic theory of the electron. This determines the various states of energy which can be occupied by an electron, that is, the values of energy which an electron can possess. The number of these states is infinite. This, however, is not a new fact because it occurs also in classical mechanics. What is new and characteristic of the relativistic theory of Dirac is that the energy of the free electron can have not only any positive value greater than or equal to mc^2 but also to any negative value which is less than or equal to $-mc^2$.

When the electron is in one of these states of negative energy both its rest energy and its kinetic energy are negative. There is a clear mathematical justification of this situation in Dirac's theory but it lies completely outside our normal way of thinking.

Thus an electron (as a result of what has been called its laziness) tends always to occupy the lowest energy states. This tendency should lead all the electrons to the negative energy states. Thus the matter which we observed could not exist.

In order to get over this fact Dirac completed his theory by adding two hypotheses. The first is that in normal conditions all the negative energy states are occupied by electrons; so that no other electron could move into them because of Pauli's Exclusion Principle which forbids a given state being occupied by more than one electron. Dirac's second hypothesis was that this sea of electrons in negative energy states could not be observed by any technique. These hypotheses may appear rather fantastic and Dirac's theory at first sight gave the impression of an elegant methematical scheme which had little to do with the physical world; but Dirac was able to deduce certain consequences of his theory which, as we have already said were soon to obtain spectacular experimental confirmation.

He showed, for example, that a photon or quantum of sufficiently high energy (greater than $2mc^2$) could in the neighbourhood of a nucleus be, in a manner of speaking, absorbed by one of these electrons in a negative energy state, which would thus acquire enough energy to transfer it into a state of positive energy. There would appear therefore one of our normal electrons which can be physically observed, but this promotion of an electron from a negative energy state leaves what has been called a 'hole'. This *absence* of a particle of negative charge and negative energy shows itself in the *presence* of a particle having the same mass as the electron but of positive electric charge and of positive energy; the hole, that is, behaves like an ordinary electron which has, however, positive electric charge; and it is this which we have called a positron. Thus we have the disappearance of a gamma quantum and the simultaneous creation of an electron-positron pair.

Anderson's discovery of the positron, that is of a particle which has the properties which two years earlier Dirac had foreseen for the holes in the negative energy states, was a spectacular confirmation of Dirac's theory. Since then there have been photographed in Wilson chambers hundreds of these processes of gamma quanta realization which as the theory has shown cannot take place in empty space but only in the immediate neighbourhood of an atomic nucleus.

The transfer of an electron from a negative energy state to a positive energy state consists in the creation of an electron-positron pair; the opposite process is the transfer of an electron from a positive energy state to a negative energy state (which exists free in the sea of negative energy electrons) and which leads to the annihilation of an electron-positron pair with the emission of energy in the form of gamma quanta.

When a gamma quantum of energy greater than $2mc^2$ falls on matter it creates an electron-positron pair and these two particles, when they move through the surrounding matter, lose energy as a result of those phenomena of excitation and ionization which we have already discussed. The electron, once it has been slowed down, moves to and fro amongst the molecules until it becomes captured by an atom which lacks one of its planetary electrons; but a positron, once stopped and sometimes even before it is stopped, disappears because an electron goes to fill the hole, i.e. to occupy the vacant state of low energy; this occurs naturally with the emission of the surplus energy. One has, therefore, the simultaneous disappearance of an electron and a positron; one has, that is, an annihilation of matter. This process explains the rarity of positrons. In a world in which positrons predominated electrons would be exceedingly rare. They would undergo the same fate which positrons undergo in the world which surrounds us.

A photon therefore can provoke the emission of an electron from the material which it crosses by three different processes: the photo-electric effect, the Compton effect and the production of electron-positron pairs. In the first and the last of these processes the incident photon is completely absorbed. In the second, on the other hand, it is scattered with a reduction of frequency. Furthermore, in the photo-electric effect and in the Compton effect the electron which is freed existed previously in the atom. It was one of its planetary electrons; on the other hand, in the third process, a new electron is created to make a pair with the positron.

We may ask what circumstances decide which of these three processes will occur? It has been found that their relative probability depends on the frequency of the incident photons and on the atomic number of the material. If the quanta have a relatively low frequency (that is, if the radiation is a beam of X-rays) electrons are effectively emitted by photo-electric effect; as one goes to higher frequencies (when the incident radiation is made up of gamma rays) the number of electrons emitted by the Compton effect increases rapidly and no further electrons are emitted by the photo-electric effect. Finally, when the frequency of the incident quanta is such that the energy is greater than $2mc^2$ (where, as usual, m is the mass of the electron and c the velocity of light) pair production begins and becomes dominant for gamma rays of very high frequency.

With the discovery of the positron the number of elementary particles was further increased. There were then four, protons, neutrons, electrons and positrons. There were beginning to be too many; but it was only the beginning . . .

NUCLEAR FORCES

I shall speak later about the accumulation of elementary particles, of the story of their discovery, of their number and of their characteristics. As we shall see, in the description of their properties, nuclear forces play an essential part. Nuclear forces are those which bind together the protons and the neutrons which make up a nucleus. We will try to explain some of the ideas on the nature of these forces.

First of all nucleons, since they have a mass, exert gravitational forces on each other just as there are gravitational forces between planets and stars; but if calculations are made applying Newton's law it is found that, because of the extremely small values of the masses, gravitational forces are extremely small despite the very small distances between the particles. Gravitational forces are certainly not adequate to account for the high binding energy of an atomic nucleus, which in, for example, the nucleus of helium has the value of 28·18 million electron volts; as the energies increase in proportion to the atomic number for the nucleus of the isotope 238 of uranium they reach the value of 1,782 million electron volts.

In addition, in every nucleus which contains more than one proton, that is in all nuclei with the exception of that of hydrogen (i.e. of helium up to uranium), there is a repulsive force which is

The Atomic Nucleus

very much stronger than the attractive force of gravitation. This is an electrostatic repulsive force which acts between the protons which all have positive electric charge; as is well known, bodies which have electric charge of the same sign repel each other. The mutual repulsion which acts between protons is so great that, according to a calculation made by the physicist, Soddy, if one could concentrate a gramme of protons on a point on the surface of the earth and another gramme on a point on the other side of the earth the repulsion between these two groups of charges would be equivalent to a force of 28 tons weight.

If, therefore, only the forces which we already know (gravitational and electrostatic) acted between the constituents of a nucleus the repulsive force would be very much greater than the attractive. The protons and the neutrons would then be dispersed and nuclei could not exist; but as we know they do exist and furthermore nucleons are bound into the nucleus so strongly that if we want to separate them one from another we must spend an energy of millions of electron volts; to be precise for every nucleus an energy equal to its binding energy.

There must therefore be in the nucleus other attractive forces which can have for stable nuclei a value high enough to overcome the electrostatic repulsive force. These forces have been called nuclear forces. They should act only over very small distances such as those which exist inside the nucleus of an atom. It has been found that the repulsive electrostatic force which acts between two protons ceases to be effective when the two particles are separated by a distance which is less than about $\frac{4}{10}$ of a billionth of a centimetre (4×10^{-13} centimetre).[1] This shows that at these distances nuclear forces prevail over electrostatic forces and that they have a radius of action of about this size.

As opposed to the forces which hold together an atom (which are electromagnetic forces) nuclear forces are of a new type which has no analogue in classical physics. It is for this reason that the theoretical study of the structure of the atomic nucleus is very much more difficult than the study of atoms and of

[1] The length 10^{-13} centimetres is often used in nuclear physics and has been given the name of *Fermi* from the name of the Italian physicist, Enrico Fermi. The diameter of a heavy nucleus is about 15 fermi; the diameter of the atom of hydrogen is about 100,000 fermi.

molecules; the difficulty in the study of atoms and molecules consists in the application of laws of force which are already well known. With the nucleus however there is not only the difficulty of applying quantum mechanics to complicated systems; one does not even know the laws which govern nuclear forces. Thus in nuclear physics there are two successive problems; first one must find the laws of nuclear forces and then calculate the nuclear properties which arise from these forces.

Up till now after twenty-five years of intensive study there is no theory of nuclear forces which is completely satisfactory. One can understand therefore why nuclear physicists, in order to account for the experimental results obtained in nuclear physics laboratories, make use of simple intuitive models each of which is an attempt to interpret a particular class of experimental results. As the physicist Weisskopf has said, 'In our attempts to penetrate into the invisible world of a nucleus we are in a situation of the three blind men who examined an elephant; different experiments supply different aspects of the objects. Our different nuclear models are not mutually exclusive; often they can be combined to allow a more complete understanding of the atomic nucleus.' Thus, although apparently contradictory, the different models should be considered as constituent parts of a whole each of which is adapted to answer certain questions on the behaviour of the nucleus.

I will deal later with models of the nucleus. We will now return to nuclear forces and will see what one can say about their intensity and their radius of action.

In a nucleus made up of protons and neutrons one can consider three different types of pairs of particles: neutron-neutron, proton-proton, neutron-proton. The first question which is naturally asked is: are the forces which act between these three pairs of particles different, or are they all the same?

The answer is that, apart from the electrostatic repulsion which is added to nuclear forces in the case of the proton-proton pair, nuclear forces are identical in the three cases. The equality of nuclear forces for the neutron-proton pair and for the proton-proton pair has been shown directly by experiment in comparing the collision of neutrons against protons and of protons against protons. Based on simple symmetry considerations (which have been fully justified indirectly by numerous experimental facts) one arrives at the conclusion that the forces which exist between two neutrons are also equal to those which exist in the other two cases. For this reason we can speak of nuclear forces without specifying every time to which of the three pairs of particles we are referring.

This result is of very great importance and was immediately put forward as a fundamental principle of physics; it is called the principle of *charge independence*. It is valid wherever one can neglect electromagnetic interactions, not only for nuclear forces but also for all those interactions which are called 'strong interactions' about which I shall speak in the last chapter.

A fundamental advance in the exploration of nuclear forces was made by the physicist, Wigner. Although nuclear forces are attractive, they cannot be in a manner of speaking completely attractive because otherwise all nuclei would collapse; that is, all the particles making up the nucleus would come into such close contact with each other that each would exist in the region of attraction of all the others. As a result, the binding energy of each nucleon would increase as the total number of particles in the nucleus increases; it should, that is, become greater and greater as one moves from the lighter nuclei to the heavier nuclei. Now this does not happen. In fact if for each nucleus one divides the value of its binding energy by the number of particles which make it up one has the mean binding energy per particle. Now this, as we shall see, is found to be almost constant for all nuclei; and furthermore it is found that the volume of a nucleus increases proportionately with the number of its constituent particles.

This shows that nuclear binding forces possess the property of *saturation*. They are forces which are in some way similar to chemical valence. To make this clearer one can think for example of a nucleus which contains a hundred particles. Saying that nuclear forces exhibit the phenomenon of saturation is equivalent to saying that each of these particles interacts only with those which are closest to it and not with all the hundred particles which make up the nucleus. This is a characteristic property of nuclear forces; electrical forces for example do not have this property of saturation.

In order to account for this saturation of nuclear forces Heisenberg in 1932 proposed a theory which was later in the same year modified by Majorana. In this theory one makes the hypothesis that the nuclear force which acts between a neutron and a proton is an exchange force (analogous to that which, for example, binds together the two atoms in a molecule of hydrogen): that is that the forces between a neutron and a proton lead to a continual exchange of their position.

The exchange occurs mainly if two nucleons move in very similar orbits and since by Pauli's Exclusion Principle (which is valid not only for planetary electrons in an atom but also for nucleons) only a few particles can exist in neighbouring orbits these exchange forces act so that each nucleon is effected by an attractive force only from a few of the others.

Experiment has confirmed this hypothesis—but only partially; In the sense that it has been found that the forces which bind nucleons together are not all of the exchange type and that there must therefore be another factor which contributes to the saturation. This almost certainly consists in a reversal of the direction of nuclear forces for extremely short distances; so that when two nucleons arrive 'too close to each other' attraction is turned into repulsion.

This idea of a 'repulsive core' for the forces is not a new concept in atomic physics. When atoms are united to form a chemical compound they are held together by attractive forces; but these act only until the atoms are in contact with each other, that is, until there occurs a significant super-position of their electron clouds. For smaller distances the exclusion principle keeps the electrons separate and acts as a strong repulsive force.

Experiment has confirmed the repulsive nature of nuclear forces at exceedingly short distances.

In conclusion, I can say that: nuclear attractive forces begin to act between two nucleons when they are separated by a distance of about four fermis; but when this distance becomes less than 0·4 fermis the attractive force becomes repulsive: two particles then repel each other.

NUCLEAR MODELS

I have explained (of course in a very rough and qualitative manner) what we think today about the character of nuclear forces. It is a theory which is not yet complete. Classical

The Atomic Nucleus

mechanics describes the motion of macroscopic bodies and quantum mechanics completely explains the behaviour of the atom; but the nucleus is a world which is only partially explored and is full of uncertainty. We do not yet know how to write mathematical expressions for nuclear forces nor the equations which control the motion of nucleons in the nucleus.

There have been many nuclear models which have been put forward to try to bring together in an organized whole the results supplied by experiments and to provide the imagination with a guide for possible prediction. The principal models are: the shell model, the drop model, and the optical model. Although they are apparently contradictory they must be considered as parts of a whole. As I have said, each is suitable for interpreting one particular part of the behaviour of the atomic nucleus.

The most obvious idea was to use for the atomic nucleus the same model which had had such notable success for the atom. According to this model, planetary electrons move into the atom around the nucleus in different orbits. They can, however, be considered as grouped in layers each of which is made up of orbits which are nearly the same. It was thought that the nucleus could be considered as made up of protons and neutrons grouped together in orbits or shells. Each of the nucleons moves independently in an orbit under the influence of a force which represents the average effect of all the other nucleons, just as an electron moving in the orbit of an atom is controlled by an average force which represents the influence exerted on it by the nucleus and the other electrons.

When an atom has a number of electrons such that a shell of orbits is complete it is particularly stable and we have an atom of a noble gas. When the first shell is complete we have helium (two planetary electrons), when the second shell is complete (ten electrons) we have neon, when the third is complete (eighteen electrons) we have argon, etc. The stability of the noble gases (which only with difficulty form ions and enter into chemical combination with other elements) is so characteristic that their atomic numbers (that is the number of their planetary electrons) deserves special designation: 2, 10, 18, 36, and 54 can be called the magic numbers of the periodic table of elements. One notes that these are all even numbers.

Well then, just like the electrons of the atom the particles making up the atomic nucleus also like certain magic numbers. If in fact one observes the proton-neutron diagram—or better, a table of isotopes of different elements—one sees that both protons and neutrons have a strong tendency to appear in even numbers. The elements in which the nucleus has an even number of protons normally possess a greater number of isotopes than those possessed by the elements with odd atomic number. The number of nuclei with an even number of neutrons is much greater than the number of nuclei which contain an odd number. The study of the abundance of elements in the earth leads to the conclusion that the elements with even atomic number make up about 87 per cent of the earth's crust. In the whole list of isotopes there are not more than six stable nuclei which contain an odd number of protons and an odd number of neutrons.

All of this indicates that the nuclei with an even number of protons and neutrons are the most stable and amongst these there are some which show themselves to be particularly stable; that is, their constituent particles are bound together with a binding energy which is much greater than the binding energy of other nuclei: helium (consisting of two protons and two neutrons), oxygen (eight neutrons and eight protons), calcium (twenty protons), nickel (twenty-eight protons), barium (eighty-two neutrons), lead (eighty-two protons), etc.

One can therefore make also for nuclei a list of magic numbers analogous to that which has been made for the atoms. They are 2, 8, 20, 28, 50, 82, 126. Nuclei with this number of protons or neutrons are exceptionally stable. The originators of this shell model think that these magic numbers represent complete shells in the nucleus analogous to the shells of electrons in the atom. The first nuclear shell would contain two nucleons, the second six, the third twelve, the fourth eight, the fifth twenty-two, the sixth thirty-two, the seventh forty-four.

This model of the nucleus, which was proposed some time ago but which has been perfected especially by the work of Maria Mayer at the University of Chicago, allows one to explain a large assembly of data on the behaviour of a nucleus so that there is now no doubt about its essential validity.

In order to overcome certain serious difficulties which were met in attempts to interpret the results of experiments on collisions of neutrons against nuclei using the original shell model (before it was perfected by Maria Mayer) Bohr in 1936 proposed the *drop model*. The fact that the nucleons in a nucleus are held together by exchange forces which act on neighbouring particles implies that a nucleon which is in the inside of a nucleus is pushed equally in all directions whilst a nucleus which is at the surface of a nucleus is attracted towards the centre. There exists therefore a surface effect which tends to make the nucleus take up the form of a sphere; but this is just what happens in a drop of liquid. For example, a drop of water is made up of an assembly of molecules of water and it has a spherical form because of surface tension. In this sense, one can say that an atomic nucleus is similar to a drop of liquid. If a nucleus is struck by a neutron coming from outside it behaves, as we shall see, in a way similar to a drop of water which has been struck by a molecule of water.

This drop model for the nucleus was extremely successful and led to an explanation of several different characteristc aspects of nuclear reactions.

Other models have been proposed for the nucleus (amongst which we should remember the optical model) and have been shown to be useful in the study of particular problems. I do not believe, however, that it would be useful for me to dwell further on this subject.

We shall now see how one can explain the emission by atomic nuclei of the gamma rays of the alpha particles (helium nuclei) and of the beta particles (electrons) which are observed in radioactive phenomena.

EMISSION OF GAMMA RAYS

Just as an atom, when it moves from one excited quantum state to a lower quantum state, emits a photon or quantum of light so a nucleus, which is in an excited state and which passes to a state of lower energy, emits a photon which has an energy so high that its frequency is higher than that of X-rays. These are the high energy photons which constitute the gamma rays emitted by radioactive substances in which the nuclei, as the result of emission of alpha or beta particles about which we shall speak soon, are often in excited states. The transition to the ground state occurs with the emission of gamma quanta.

One understands then that the emission of gamma quanta by radioactive nuclei is always a secondary process. The primary process is the emission of alpha particles or beta particles.

ALPHA DISINTEGRATION

Alpha particles which are emitted by radioactive nuclei already exist as such inside the nucleus. Their existence in the nucleus certainly does not conflict with the concept of the nucleus being made up of protons and neutrons; for an alpha particle (mass 4 charge 2) is nothing other than a nucleus of helium made up of 2 neutrons and 2 protons. The fact therefore that an element emits alpha particles shows only that in the nucleus some protons and some neutrons are bound together closely to form alpha particles.

However, if one compares the phenomenon of the spontaneous emission of an alpha particle by a nucleus with the phenomenon of the collision of an alpha particle against the same nucleus one encounters a difficulty.

When a particle which arrives from outside with a certain energy comes close to a nucleus it is repelled by a force which becomes continually greater as the distance between the particle and the nucleus decreases (for both particle and nucleus are positively charged). But when it reaches a distance from the centre of the nucleus equal to the nuclear radius R the force changes sign abruptly; that is, at the moment when the particle enters the nucleus the repulsive force becomes attractive. The behaviour of this force is shown schematically in Fig. 29; the abscissa represents the distance, r, of the particle from the centre of the nucleus and the ordinate the value of the force, F, which acts between the nucleus and the alpha particle; R represents the radius of the nucleus.

If therefore an alpha particle which is inside the nucleus wants to get out it will succeed only if it has an energy sufficiently high to overcome the attractive forces which act on it when it is inside the nucleus; but as soon as it has passed this 'barrier' which holds it inside it will be repulsed with a force which is initially very large but which diminishes as the particle moves away according to Coulomb's law of electrostatic repulsion.

According to classical theory, an alpha particle which is inside an atomic nucleus, could pass the barrier only if it had an energy greater than that corresponding to the height of the barrier; but this conclusion is not confirmed by experiment. It is in fact found, for example, that a nucleus of uranium when it

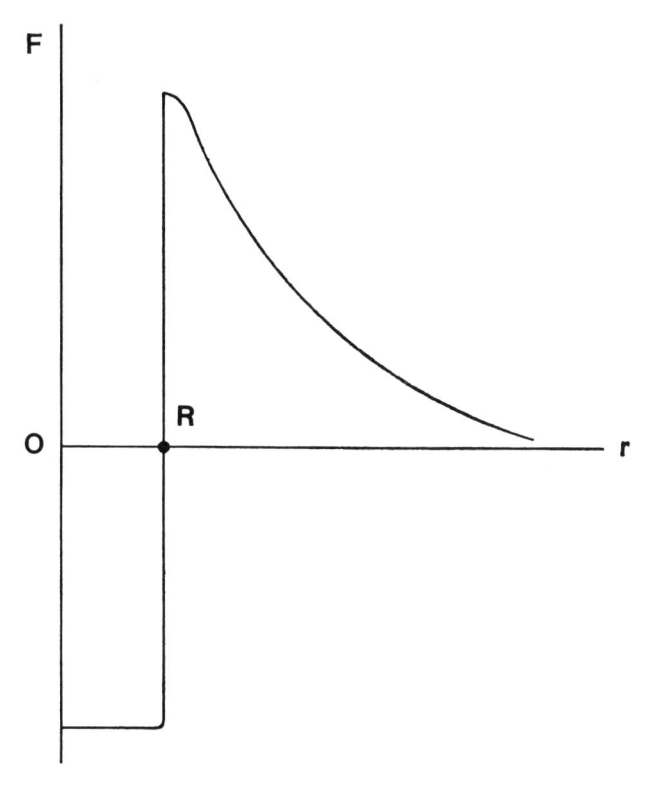

FIG. 29. The nuclear force.

disintegrates emits alpha particles whose energy is $6 \cdot 6 \times 10^{-6}$ ergs whilst the energy corresponding to the height of the barrier is 14×10^{-6} ergs, that is about double the energy of the alpha particle.

In 1928 Gamow overcame this difficulty by treating the problem from the point of view of wave mechanics. According to wave mechanics, even in the case in which the alpha particle has an energy less than that which is necessary to overcome the barrier it has a certain probability of escaping from the nucleus. This probability is very small if the energy of the alpha particle is very much less than that necessary to overcome the attractive forces; but it increases as its energy comes close to this value

Starting from these ideas Gamow was able to give a satisfactory explanation of most of the phenomena relating to the emission of alpha particles. His theory also accounted perfectly for the enormous variation in the mean lives of radioactive substances which, as we have seen, varies from minute fractions of a second to many millions of years; Gamow's theory also explains why substances with shorter mean lives emit particles of higher energy.

BETA DISINTEGRATION

We will now consider the emission by radioactive nuclei of beta particles, that is of electrons.

We know that a nucleus is made up of protons and neutrons and that it does not contain electrons; but nevertheless experiment showed that electrons are emitted from the nucleus. How can we explain this paradox? How, that is, can we explain the fact that a nucleus emits particles which it doesn't contain?

One encounters a similar paradox in the study of the atom: the problem of the emission of light by atoms. Electromagnetic radiation does not appear to play a part in atomic structure but nevertheless electromagnetic radiation emerges from atoms. This contradiction was resolved by making the hypothesis that the atom can change its state (passing from a higher energy state to a lower energy state) and simultaneously emit a light quantum. Thus although electromagnetic radiation has no part in the real structure of the atom it does have a type of latent existence in atoms.

It would seem, therefore, natural to regard the emission of electrons by atomic nuclei in a similar way; but there is a serious difficulty. This process seems to contradict a fundamental law of physics—the law of conservation of energy (in which, of course, is included the so-called *rest energy* of the various particles, the energy equivalent to their mass). When an atom moves from a higher energy state to a lower energy state it emits a quantum of well-defined energy which is equal to the energy liberated in the atomic transition. One would expect that a similar process occurs in the emission of electrons by nuclei. Let us for example consider a certain number of atoms of uranium X_1 which disintegrate, emitting beta particles, and are transformed into uranium X_2. As the initial nuclei and the final nuclei have fixed energies the electrons which are emitted should all have the same energy equal to the difference in energy between the nuclei before and after the emission; but it is found that the electrons which are emitted have all possible values of energy from zero up to a maximum value of 780,000 electron volts.

Are we then faced with an exception to the fundamental law of conservation of energy?

THE NEUTRINO

The discovery that the electrons emitted by nuclei have all possible values of energy up to a maximum value was made by Chadwick in 1914. It did not, however, immediately attract attention. It was thought that the slower electrons were those which came from atoms in lower layers of the radioactive material; in other words, it was thought that all electrons were emitted with the same energy but that those electrons which came from atoms which were inside the material had lost part of their energy in crossing the material before they were able to escape to the outside.

And for a good thirteen years this simple explanation was accepted by everyone. It was accepted until delicate experiments in which the total heat liberated in a radioactive process was measured showed that the electrons emitted from disintegrating atoms lose hardly any energy inside the radioactive material but are effectively emitted from the nuclei with very different energies.

The German physicist, Pauli, then put forward a hypothesis which was able to explain the phenomenon of the emission of electrons by radioactive substances and which rescued the hallowed principle of the conservation of energy. He assumed the existence of a new elementary particle which should be electrically neutral and which should have a mass even smaller than the mass of the electron; it should, moreover, have a high penetrating power so that it would be able to pass very close to other particles without being captured by them.

As this particle is electrically neutral Pauli at first called it the neutron but, as we know, Chadwick in 1932 discovered a new particle which was electrically neutral and whose mass is equal to that of the mass of the proton and it was this particle which was officially given the name of neutron. One day when the Italian physicist, Fermi, in a lecture at the University of Rome referred to this discovery a student asked if Chadwick's neutron was the same as the neutron which had been proposed by Pauli to account for the phenomenon of the emission of electrons by radioactive substances. 'No,' replied Fermi, 'Pauli's neutron is very much smaller. It is a neutrino.' This name, which was introduced as a joke, has stuck and is now universally used to indicate the particle which was proposed by Pauli.

We will now see how Pauli's hypothesis of the existence of a neutrino accounts very well for the fact, which had appeared so strange, that the electrons which are emitted by radioactive substances have energies of all possible values up to a maximum value. Let us suppose that a nucleus of uranium X_1 emits simultaneously an electron and a neutrino. Then the difference in energy between that of the initial nucleus of uranium X_1 and of the final nucleus of uranium X_2 will be divided between the electron and the neutrino; it can be divided in any way so that we shall have many pairs of electrons and of neutrinos with all possible energies provided only that the sum of the energies of the neutrino and of the electron always has the same value; this value will be equal to the difference in the energies of the initial nucleus and the final nucleus. Now given that the neutrinos, as a result of their lack of electric charge and of their very small mass, escape from every method of observation we will only observe as the emission products of uranium X_1 electrons of all possible energies. The maximum energy which an electron can have will then correspond to the case in which the electron is emitted with all the energy without sharing any with the neutrino. It is in fact found that this maximum energy has just the value given by the difference in energy between the initial nucleus and the final nucleus.

The Atomic Nucleus

This was how this new particle was introduced into physics. It is a strange particle which has no mass, no electric charge and virtually possesses only one property, that of allowing physicists to maintain their unshakable faith in the principle of the conservation of energy. Without the neutrino beta-disintegration could not be reconciled with the law of conservation of energy.

Naturally experimental physicists have tried with all methods to check by experiment the existence of this new hypothetical particle; but for years it escaped from all attempts to detect it. The neutrino interacts exceedingly rarely with an atomic nucleus; it would on the average travel for four light years in lead (that is travel about 40 billion kilometres) before interacting with a nucleus. Neutrinos appear to take no account of the matter which they traverse. How can one then hope to surprise one in the act of 'doing something'. It wasn't until June of 1956 that the neutrino was detected by Cowan and Reines of the laboratory of Los Alamos. They used for their experiment (a brilliant experiment on an epic scale) an ingenious arrangement which made use of the powerful nuclear reactor of Savannah River.[1]

FERMI'S THEORY

Basing himself on Pauli's hypothesis of the existence of the neutrino the Italian, Enrico Fermi, constructed in 1934 a theory which resolves the problem of the emission of electrons by atomic nuclei; i.e. he produced a theory which allows one to reply to the question: how is it possible that a nucleus which is made only of protons and neutrons can emit electrons?

The apparent conflict between beta decay and the law of conservation of energy was resolved by the hypothesis that radioactive nuclei emit simultaneously electrons and neutrinos. One can therefore take up again the analogy between the emission of electrons by nuclei and the emission of gamma quanta on the part of atoms. A quantum is emitted by an atom when one of its planetary electrons moves from one orbit to another of lower energy, that is when an atom changes from

---

[1] About ten per cent of the energy of an atomic pile is emitted in the form of neutrinos each of which has an energy of a few Mev.

one quantum state to another of lower energy. Fermi in his theory supposed that the neutron and the proton are only two different quantum states of the same heavy particle, the nucleon. When the nucleon passes from the neutron state to the proton state an electron-neutrino pair is emitted even though this did not previously exist in the nucleus.

Inside a nucleus then a neutron can change itself into a proton; but this can also happen to a free neutron which spontaneously transforms itself into a proton and an electron. The mean life of a free neutron is about twelve minutes; that is, a free neutron lives on average twelve minutes before changing itself into a proton and an electron. Despite this one can never succeed in observing the free neutron in matter because in an exceedingly short time it reacts with any atomic nucleus. If one were able, for example, to withdraw 1,000 neutrons from any interaction with the external world, after twelve minutes 500 of them would have changed themselves into protons. After twenty-four minutes there would remain only 250 and so on.

As we shall see later, in certain cases positrons (that is positive electrons) are emitted by radioactive substances. According to Fermi's theory this emission of positrons could be interpreted as due to a process inverse to that previously considered, that is to the transformation of a proton into a neutron.

Using this theory one can explain many of the characteristics of the emission of beta rays by atomic nuclei.

THE HYPOTHESIS OF THE EXISTENCE OF THE MESON

We have seen that in Fermi's theory of beta disintegration the proton and the neutron are considered as two different states of the same body, the nucleon. The change from the neutron state to the proton state is accompanied by the emission of an electron and a neutrino, whilst the inverse change is accompanied by the emission of a positron and of a neutrino.

There arose therefore the idea that such a theory might not only explain the radioactive emission of electrons and positrons but also contain as a natural consequence the existence of the exchange forces between protons and neutrons which had been introduced in an *ad hoc* hypothesis by Heisenberg and Majorana.

After, following Maxwell, the old idea of action at a distance was abandoned every time that one speaks of an interaction between two particles one assumes the existence of a field. Thus, two electrically charged particles exert forces on each other by means of the electromagnetic field. Now the quantum theory tells us that the electromagnetic field in which disturbances are propagated as waves should also have a corpuscular aspect which is represented by photons. 'Wave mechanics and then the quantum theory of fields', as has been put very clearly by the French physicist, Leprince Ringuet, 'has led to a precise definition of the mechanism by which interactions can be transmitted between particles. Each of two electrons is in interaction with the electromagnetic field in which it is immersed. This corresponds in corpuscular language to the possibility that an electron can emit a photon which the other will absorb. To speak of the interaction between two electrons or of the possible emission and absorption of photons is the same thing. This idea of the particle connected to a field is fundamental and completely general.'

A bold application of these ideas to the study of nuclear forces was made in 1935 by the twenty-eight-year-old Japanese physicist, Yukawa, who was awarded the Nobel Prize in 1949. As interactions take place between the nucleons which make up an atomic nucleus it is logical to think that these occur by means of a field. We know, however, that this cannot be the electromagnetic field nor the gravitational field for nuclear forces are neither electromagnetic nor gravitational. It will be a new type of field which will, however, have like the others its corpuscular aspect, that is, there will be associated with it particles (analogous for example to the quanta of the electromagnetic field) which will be emitted and absorbed in interactions between the nucleons. He introduced a formula by means of which one can calculate the mass of the particle (i.e. the corpuscular aspect of the field) exchanged between two bodies when the range of their interaction force is known. Yukawa found that the mass of the particle-quantum of this new nuclear field, which is emitted and absorbed by nucleons during an interaction between them, should be about three hundred times greater than the mass of the electron.

As this new particle has a mass intermediate between the

mass of the electron and that of the proton, Yukawa called it the *meson*.

Mesons are therefore the quanta of the nuclear field which can for this reason also be called the meson field. They have been called 'the nuclear cement which binds together the protons and the neutrons in the nucleus'.

Since there exist in the nucleus three types of binding which are equally strong (neutron-neutron, proton-proton, neutron-proton) it was supposed that there are three types of mesons, that is the meson with positive electric charge (for the proton-neutron interaction), the negative meson (neutron-proton) and the electrically neutral meson (proton-proton and neutron-neutron).

Yukawa's hypothesis, although interesting, appeared at first sight very artificial for it assumed the existence of new particles (two with opposite charge and a third with no charge) having masses intermediate between those of the electron and of the proton. The hypothesis appeared even more artificial because Yukawa, in order to be able to explain radioactive emission of electrons and positrons, attributed to these particles another strange property. He supposed that the charged mesons are unstable with a half life of about 10^{-6} or 10^{-7} seconds and that they gave rise to an electron (positive or negative according to their charge) and a neutrino, destroying themselves in the process. He supposed that the neutral particles would have an exceedingly short mean life and should be transformed into gamma quanta.

As one can see, this was a collection of extremely bold hypotheses; so for some time Yukawa's ideas were treated as pure and simple speculation. We shall see later that certain experimental facts which appeared in the study of the cosmic radiation gave birth to the suspicion that in these speculations there were elements of truth.

CHAPTER SEVEN

ARTIFICIAL RADIOACTIVITY

ARTIFICIAL TRANSMUTATIONS

Rutherford's experiment had shown that nitrogen when bombarded by alpha particles emitted protons but what happened to the bombarding alpha particle? What happened to the nucleus which was bombarded?

The experiment was repeated in a Wilson Chamber and examination of the tracks on the photographs confirmed what Rutherford had said: that nitrogen nuclei absorb alpha particles and emit protons thereby transforming themselves into oxygen nuclei.

Finally, therefore, nitrogen has been changed into oxygen. For the first time man had succeeded in provoking an artificial transmutation. He had succeeded that is in changing one element into another.

After this first experiment by Rutherford the forms of atomic transmutations were enormously extended and refined. Nowadays one can operate on nuclei and provoke their transmutation with any agent which has sufficient energy: alpha particles, protons, deuterons, neutrons and gamma rays.

There is however a serious disadvantage to bombardment by charged particles (alpha particles, protons, deuterons). If a charged particle is to reach a nucleus it must overcome the potential barrier due to its electric charge, as I have explained in the previous chapter, and the greater the charge of the nucleus, that is the greater its atomic number, the greater is this barrier. Thus a charged particle projectile will succeed in breaking up a nucleus only if it has enough energy to overcome this nuclear potential barrier.

The alpha particles used by Rutherford were of rather low energy which was adequate only to overcome the potential barrier of the lighter nuclei. With this method it is possible to disintegrate only the elements whose atomic number is less than 30 (zinc). If one wants to disintegrate the heavier atoms by means of bombardment with charged particles one has to use particles of higher energy. Nowadays these particles can be obtained with special machines called accelerating machines; I will later devote a whole chapter to them.

But in the list of projectiles at the disposal of physicists there is one which has no electric charge, the neutron. Because of this absence of charge a neutron will not be influenced by the electric field of the nucleus (Fig. 30) and will not, therefore, notice any

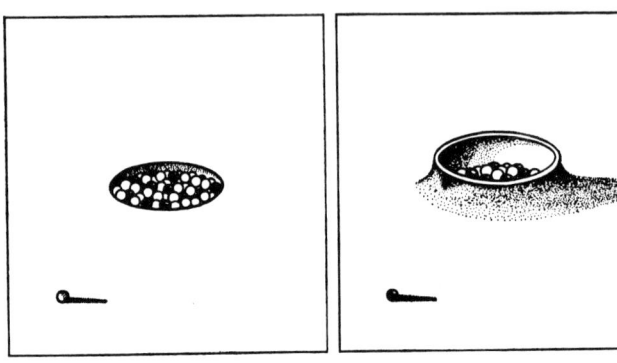

FIG. 30. Whilst a neutron is not affected by the electric field of a nucleus, a charged particle, to enter a nucleus, must overcome the potential barrier due to its electric charge.

difference between a light nucleus and a heavy nucleus. It can be used with the same probability of success to bombard a nucleus of hydrogen and one of uranium.

Thus for energies which are not very high neutron projectiles are preferable to electrically charged particles. On the other hand, for sufficiently high energies one can carry out bombardments with particles of all types, neutrons and charged particles.

NUCLEAR REACTIONS

The reactions which take place between the nuclei can be represented in shorthand by an extremely simple symbolism.

Artificial Radioactivity

We should remember that in order to represent nuclei one uses the same letters which one uses to indicate atoms and one puts at their side two numbers. One on top, and on the right indicates the atomic weight and another below, and on the left indicates the nuclear charge of the element. Thus, for example, the symbol $_7N^{14}$ indicates the nucleus nitrogen with atomic weight 14 and atomic number 7.

To represent a nuclear reaction one first writes the symbol of the bombarded element. One then puts in brackets first the symbol of the colliding particle and then that of the particle which is expelled and finally outside the bracket one writes the symbol of the residual nucleus. For example the phenomenon which occurs when nitrogen is bombarded with alpha particles is indicated by the symbol

$$_7N^{14} (\alpha, p)\ _8O^{17}$$

which means that when an alpha particle collides with a nitrogen nucleus a proton is expelled and a nucleus of oxygen remains.

This reaction can also be written

$$_7N^{14} + \alpha \rightarrow\ _8O^{17} + p$$

or

$$_7N^{14} +\ _2He^4 \rightarrow\ _8O^{17} +\ _1H^1$$

VARIOUS TYPES OF TRANSMUTATION

We will now look at some examples of various types of transmutation.

The physicists Cockcroft and Walton were the first to effect transmutations using as projectiles high energy protons, that is protons with a high velocity. The first experiment was carried out on lithium. It showed that a nucleus of lithium after having absorbed a proton projectile explodes and breaks into two fragments. These fragments are alpha particles. The process occurs according to the reaction

$$_3Li^7 + p \rightarrow\ _2He^4 +\ _2He^4$$

and so lithium is transformed to helium.

Nuclear disintegration produced by means of bombardment with deuterons (D) were first studied by Lawrence and his collaborators. One case of this bombardment is particularly interesting, the bombardment of deuterons at rest by high energy deuterons. This experiment showed that in this case there may occur two reactions. Sometimes the nucleus of heavy hydrogen which is bombarded after having absorbed the projectile deuteron breaks up into a proton (that is a nucleus of ordinary hydrogen) and into a nucleus of an isotope of hydrogen which is even heavier than deuterium, that is tritium. As I have said before a nucleus of normal hydrogen is made up of one proton; a nucleus of deuterium (that is a deuteron) is made up of a proton and a neutron; a nucleus of tritium is made up of a proton and two neutrons. This first process then occurs according to the reaction

$$_1H^2 \, (D, p) \, _1H^3$$

or alternatively

$$_1H^2 + {}_1H^2 \rightarrow {}_1H^1 + {}_1H^3$$

But on other occasions the bombarded deuteron may absorb the projectile deuteron and explode giving rise to an isotope of helium and a neutron according to the reaction.

$$_1H^2 \, (D, n) \, _2He^3$$

Another type of reaction is that discovered in 1934 by Chadwick and Goldharbour who were the first to bombard nuclei not with particles but with *gamma rays*, that is with electromagnetic radiation. These experiments gave positive results. We find a phenomenon analogous to the photoelectric effect and for this reason it has been called the nuclear photoelectric effect. The nuclei which were submitted to bombardment with gamma rays were deuterons; they absorbed the gamma quanta and emitted neutrons, thus transforming themselves into nuclei of hydrogen according to the reaction

$$_1H^2 + \text{gamma} \rightarrow {}_1H^1 + n$$

For these experiments one can use either gamma rays spontaneously emitted from thorium C'' or high energy gamma rays produced artificially.

ARTIFICIAL RADIOACTIVITY

In the artificial transmutations which were made up to the

Artificial Radioactivity

last months of 1933 it was always observed that a nucleus when struck by a projectile was immediately modified; it absorbed the projectile and broke up to give rise to another stable nucleus; a characteristic of the transmutation is then that its final product is stable. When, for example, a nucleus of aluminium is struck by an alpha particle it emits a proton and changes into a stable nucleus of silicon.

Towards the end of 1933 the Joliot-Curies were studying just this reaction in a Wilson chamber. They examined the transformation of aluminium bombarded with alpha particles coming from a polonium source and they examined the tracks of the fast protons which were emitted. In doing this they discovered that alongside the proton tracks there were also tracks of positive electrons coming from the aluminium. They were thus confronted with a new fact; for in no artificial disintegration had there ever been observed the emission of positive or negative electrons.

Continuing these experiments the Joliot-Curies discovered by chance that even after the alpha particle source had been removed the aluminium continued for some minutes to emit positrons. The alpha particles were therefore producing from nuclei of aluminium other unstable nuclei which could exist for some minutes but which, later, underwent a second transformation in which they emitted positrons, thus becoming stable. Artificial radioactivity had been discovered.

Thus we can already see from this experiment that a product of a nuclear transformation is not necessarily stable. It may, in certain cases, be an unstable nucleus which does not exist in nature and which, sooner or later, must undergo further modification. It emits, in fact, a particle thus transforming itself into a stable nucleus. Its behaviour is then analogous to that of natural radioactive nuclei. Furthermore, the laws by which unstable nuclei disintegrate are the same both for natural radioactive elements and for artificial radioactive elements.

In artificial radioactivity then the nuclear transformations occur at two distinct times. At the first, a projectile collides with the nucleus and immediately provokes a transformation into another unstable radioactive nucleus; then, later, this unstable nucleus spontaneously emits a particle and is transformed into a stable nucleus.

The experiments of the Joliot-Curies were not, of course, limited to aluminium. They devoted themselves to a detailed study bombarding with alpha particles other light elements such as boron and magnesium. They observed that these, too, like aluminium, became radioactive. For example, boron at first captured the bombarding particle and emitted a neutron producing in this way an unstable nucleus which has the same charge as that of normal nuclei of nitrogen, of which it is an isotope. At a later time, the nucleus of radioactive nitrogen emitted a positron and was changed into stable carbon.

Despite the very small number of atoms which are made in these transformations it has been possible to verify by chemical analysis that the radioactive nuclei formed from boron and from magnesium are in fact nuclei of nitrogen and of silicon. I will illustrate soon the method which is used to carry out the analysis.

In this way, the Joliot-Curies were able to verify that the activity of the radioactive nuclei formed from boron, from aluminium and from magnesium decays in time in the same way as that of natural radioactive elements. They found that radioactive nitrogen which is made from boron has a half life of fourteen minutes, i.e. that after fourteen minutes its activity is reduced to a half. The activity of radioactive silicon which comes from magnesium is reduced to a half in two and a half minutes and the half life of the nuclei which are obtained by irradiating aluminium is three minutes.

The half life is a value which has a much greater importance than might appear at first sight; for it is always the same for the same nucleus and is therefore a constant which can be used to identify it.

Furthermore a substance which is irradiated shows its greatest activity when it has been bombarded for a time of the order of magnitude of that in which the radioactive element which is made from it is reduced to a half. Thus in order for boron to reach its maximum activity it is necessary to expose it to the action of alpha particles for about a quarter of an hour because nitrogen which has originated from the boron has a half life of fourteen minutes.

As a result of the experiments of the Joliot-Curies other physicists took up the study of the same phenomenon and, again using alpha particles as projectiles, confirmed their results. Experiments were then started in which nuclei of various elements were bombarded with projectiles other than alpha particles. There were discovered a very large number of examples of radioactivity provoked by bombardment with protons and even more by bombardment with deuterons.

We have already seen in artificial transmutations how elements can be bombarded not only with material particles (alpha particles, protons, deuterons) but also with electromagnetic radiation, gamma rays; and we have seen how the elements bombarded with gamma rays undergo the nuclear photo-electric effect, that is, they disintegrate emitting neutrons. The German physicists Bothe and Gentner on the other hand were able to produce by bombardment with very high energy gamma rays (seventeen million electron volts) artificial radioactivity of the bombarded nuclei. For example, phosphorus when bombarded with gamma rays emits a neutron and is transformed into a radioactive isotope of phosphorus itself; this is unstable and disintegrates in its turn emitting a positron with a half life of 131 seconds. Many cases of this type of reaction are now known.

In the course of their experiments on bombarding with gamma rays Bothe and Gentner discovered an example of disintegration which allowed them to confirm a phenomenon which had already been observed by O. Hahn in the course of his researches on the disintegration of uranium.

They found that it sometimes happens that the same nucleus which has been made radioactive disintegrates with two different half lives. For example, bromine of atomic weight 80 (which is a radioactive isotope of stable bromine) disintegrates both with a half life of 18·5 minutes and with a half life of 4·54 hours. This occurs as if bromine of atomic weight 80 were really a mixture of two different types of bromine (or as they are called two isomers of bromine) which have the same atomic number, the same atomic weight (and whose nuclei therefore have the same number of protons and neutrons) but differ only in their radioactive properties.

Many examples of nuclear isomers are now known. This is one of the phenomena which has aroused greatest interest amongst physicists because the explanation of the existence of two different nuclei (although made up of the same number of protons and neutrons) is linked with certain important laws of nuclear physics. Although the explanation suggested by Weiszächer is extremely interesting I do not think that it is appropriate here to tackle the examination of the conditions in which two nuclear isomers can exist.

We shall now see how artificial radioactive substances can be identified chemically. To make the chemical analysis one would need in this case a special technique; for the atoms of radioactive substances are formed in such small quantities that no chemical reaction could reveal their presence. The device which is used to separate this very small number of radioactive atoms is based on a property which might be called team spirit. We can be certain that if an infantryman saw various groups of soldiers belonging to various arms he would approach from *esprit de corps* the group of soldiers from the infantry. Thus if, for example, an atom of radioactive nitrogen finds itself in a solution in which there exist other atoms of stable nitrogen it unites with them and shares their destiny.

In the chemical analysis of a substance which has been made active by bombardment one takes account of the fact that the radioactive element which is formed must have an atomic number close to that of the original element. One therefore proceeds in the following way. The radioactive substance is put into solution and to this is added small quantities of all the substances which in Mendeleev's Table lie close to that under examination. One can then separate by ordinary chemical means the different elements which are in the solution and by bringing them one after another close to a counter recognize which element has drawn with it the radioactivity. One can then conclude that if a certain element has drawn with it the radioactivity this implies that the radioactive atoms which have been formed are isotopes of this element.

The technique of chemical separation of artificial radioactive elements has reached today a high degree of perfection.

BOMBARDMENT BY NEUTRONS

It was the Italian Enrico Fermi who had the idea of using neutrons for this type of experiment. He was awarded the Nobel Prize in 1938 for work of which we shall now speak. The experiments which were carried out in the Physic Institute of Rome by Fermi himself and by the physicists Amaldi, Pontecorvo, Rasetti, Segré, and the chemist D'Agostino excited immediately great attention; for whilst previously artificial radioactivity had been produced only in certain light elements it was possible by experiments with neutrons to produce a very large number of radioactive elements, and heavy elements could be produced as easily as light ones. As we already know, this is due to the lack of charge of the neutron. The neutron source was a mixture of radium emanation and beryllium powder placed in a simple glass tube a centimetre and a half long. As we know, a nucleus of beryllium emits a neutron when it absorbs an alpha projectile emitted by the radium emanation. It was these neutrons emitted in the transmutation which were used for the bombardment.

In the course of their experiments Fermi and his collaborators discovered that various nuclei when bombarded with neutrons do not always disintegrate in the same way; but, according to the circumstances, the disintegration took place according to three different processes: the nucleus which is bombarded

(1) absorbs the neutron and emits a proton: reaction (n, p);
(2) absorbs the neutron and emits an alpha particle: reaction (n, α);
(3) absorbs the neutron without emitting any particle but giving off the excess energy in the form of gamma rays: reaction (n, γ).

To these three ways of disintegration there was very soon added a fourth, discovered in 1936 by the Dutch physicist Heyn: the nucleus which was bombarded captured the incident neutron and emitted two neutrons thus transforming itself into an unstable nucleus (reaction $(n, 2n)$: this then disintegrated emitting a positive electron and transforming itself finally to a stable nucleus. This process is observed, for example, in copper. In order that this type of reaction should take place it is however necessary that the incident neutron should have a very high energy—of the order of ten million electron volts.

SLOW NEUTRONS

In the course of their experiments the Rome physicists realized already by 1934 that radioactive silver does not always have the same activity. Sometimes, for example, it might emit a hundred electrons a minute and at other times eighty; after a series of observations they reached the strange conclusion that the activity changed as one changed the object which surrounded the neutron source and the silver during the time of the bombardment. In particular they noted that the activity of the radioactive silver became about thirty times greater when the neutron source and the bombarded silver were surrounded either by water or by paraffin. As the element which is common to these two substances is hydrogen they reached the conclusion that the reason for the variation in activity was associated with the presence of hydrogen near the element and the source which emitted the projectiles.

As the result of a large number of experiments it was possible to explain this phenomenon. The projectile neutrons had the same mass as the protons, that is as the nuclei of hydrogen which existed in the substance which surrounded the source and the silver during the bombardment. As soon as these neutrons were emitted they collided against the protons even before colliding against the nuclei of silver. After these collisions the protons, which at first had been almost at rest, were set in motion at the expense of the energy of the colliding neutrons. After the first collision a neutron would continue on its path in a direction and with a velocity different from that which it had at first until as the result of a second collision it would again change velocity and direction of movement and so on. When the calculations were made it was discovered that a neutron which moved through a block of paraffin or through water undergoes such a large number of collisions against protons that it finally has a very low velocity.

The neutrons which were slowed down in this process were called slow neutrons; and they exhibited properties which were very different from neutrons of higher energy.

What is the energy of slow neutrons? By a simple argument one can show that neutrons continue to collide against protons and thus continue to lose energy until they have reached the value of thermal energy; for this reason, the slowest of slow neutrons are called thermal neutrons.

Now it is known that the atoms in every body, and thus in

particular the atoms in paraffin, are in continual motion. In solid bodies this motion consists in oscillations around a fixed position with an energy, at the temperature of twenty degrees centigrade, of a fortieth of an electron volt. As long as the colliding neutrons have high energy they lose a considerable fraction of their energy in every collision against protons and therefore slow down. When, however, their energy is reduced so that it is of the same order of magnitude as the energy possessed by the protons from thermal agitation the slowing down process stops; indeed in some cases it may happen that the proton yields a little of its energy to the neutron which collides with it.

Thus, if we have, for example, a source which emits neutrons of five million electron volts and we place it in a block of paraffin there will be at any time neutrons of all possible energies, starting from that at which they are emitted down to thermal energies, that is, some tens of electron volts.

Neutrons can also be slowed down in substances other than those containing normal hydrogen. For example, in substances containing heavy hydrogen, beryllium or carbon; these are, however, less efficient. Substances which are used to slow down neutrons are called moderators.

It was recognized as the result of a number of experiments that slow neutrons are very much more efficient than high energy neutrons for the production of radioactive elements by simple neutron capture.

The Italians Amaldi and Fermi, starting from some observations made by the English physicists Moon and Tillman, were able in 1935 to bring to light a fact which for some time appeared extremely strange.

Let us fix our attention on an example. When atoms of gold capture neutrons they give rise to a radioactive isotope with a half life of 2·7 days. Now gold is very much more sensitive to slow neutrons than to fast ones; there is a twenty-fold increase in its activity if, in the course of radiation, it is surrounded by water or paraffin.

The new fact consists in this: a gold nucleus not only captures with avidity neutrons of thermal energy (those having an energy of about one-fortieth of an electron volt) but it captures with even greater avidity neutrons which have an energy between

4·87 and 5·05 electron volts. In other words, gold is extremely sensitive in a very restricted energy interval (4·87–5·05 electron volts) whilst it is only moderately sensitive to neutrons which have an energy slightly greater than 5·05 electron volts or slightly less than 4·87 electron volts.

It was quickly recognized that just as in the case of gold in almost all cases of radioactivity (n, γ) there are energy intervals, very narrow, in which neutron capture is extremely probable.

The phenomenon is very analogous to the absorption of light by a vapour, for example by sodium vapour. If, in fact, we send light of all possible frequencies through sodium vapour the vapour strongly absorbs radiations of certain well-defined frequencies which are called the absorption lines of sodium. In the case of neutrons we have then an analogous phenomenon. For example, a gold nucleus if struck by neutrons of all possible energies will absorb only those whose energy falls in a certain interval. This interval, because of the analogy with optics, was given the name of the *neutron absorption line* for a gold nucleus.

It was then discovered that in some cases there are several absorption lines separated from each other by some tens of electron volts. The existence of these absorption lines separated by such short intervals appeared in 1936 to be very strange.

Now we know what happens when light is absorbed by a sodium atom. Referring to the Bohr-Sommerfeld atomic model the absorption of light induces the transfer of a peripheral electron in the electron cloud which surrounds the nucleus from one orbit to another more external orbit whilst all the other electrons continue unperturbed.

Until 1936 it was thought that something similar happened in the nucleus. It was thought that the first shell model of a nucleus was valid and thus that in neutron capture the complicated motion of the protons and the neutrons in the bombarded nucleus was not in any way disturbed. The last neutron to arrive would simply establish itself in the lowest free energy state under the action of the forces exerted on it by the other particles in the nucleus.

But it was shown by simple calculation that although in this nuclear model neutron absorption lines are possible they should be separated from each other by some millions of electron volts; such an interval is very much higher than the intervals found experimentally. Furthermore, the width of the absorption line should also be vastly greater than that observed experimentally.

These difficulties were overcome when Bohr proposed his drop model for the nucleus. In this model nuclear particles can be represented as balls which move continually at the bottom of a hole with rather steep walls and which remain in continual contact with each other. If then, one throws a neutron against a nucleus things go in the same way as if we threw on to the ground in the direction of the hole a new ball. This ball when it fell into the hole would collide successively with the balls that were already there so that it would lose a large amount of its energy; so the energy of a neutron would be distributed amongst a large number of nucleons and in some cases amongst all of the particles already present in the nucleus. One has, therefore, as an immediate effect of the bombardment an increase in the energy of all the particles which make up the nucleus which will therefore be in an excited state.

We should note that, although in an excited atom it is only the outmost electron which acquires an excess over its normal energy, in an excited nucleus all or almost all of the constituent particles contribute directly to the phenomenon. In the excited atom the excess energy is emitted after a very short time in the form of light when the outermost electron returns from an excited orbit to its normal orbit. In the case of the nucleus on the other hand things are more complicated because various types of process are possible.

It may sometimes happen that amongst the moving particles which make up the nucleus there is one which lies in the peripheral zone and which has a high enough energy to climb up the walls of the hole. It then escapes from the nucleus and carries with it a large part of the energy of the excited nucleus. We then have one of the processes in which the capture of the neutron is accompanied by the emission of a proton or an alpha particle or by one or two neutrons. It was found as the result of certain simple calculations that there is a reasonable probability of such phenomena only if the energy of the incident particle is very high.

In other cases before the energy acquired by the nucleus as a result of the capture of the incident particles is concentrated on to one particle the nucleus moves from the excited state to the ground state by emitting the energy excess in the form of one or more gamma quanta. The process then becomes merely the simple capture of the incident neutron. This type of process normally occurs when the incident neutron has very low energy (a slow neutron).

We can therefore say, in accordance with the experimental results, that: when a radioactive element is formed by the simple process of neutron capture by the nucleus, slow neutrons (of low energy) are more efficient than fast neutrons; but in other processes (neutron capture with emission of a proton or of an alpha particle or of two neutrons) slow neutrons do not have sufficient energy to produce the disintegration of the nucleus which can only be achieved by neutrons of high energy.

Bohr showed that according to his nuclear model the process of slow neutron capture exhibits absorption lines which are characteristic of the bombarded nucleus; the width and separation of these lines agree with the experimental results.

RADIOACTIVITY PRODUCED IN HEAVY ELEMENTS

I will now dwell upon certain particularly interesting cases of artificial radioactivity; these are the cases of uranium, of protoactinium and of borium, that is, the last three elements in Mendeleev's periodic system.

Fermi and his collaborators, making a systematic study of the radioactivity produced by neutrons, realized in 1934 that even uranium, which is naturally radioactive (it emits alpha particles), gives rise under neutron bombardment to many new bodies which emit beta particles. These physicists directed their attention in particular to two bodies which have half lives of fifteen minutes and of a hundred minutes. From a systematic study of the chemical properties of these two bodies they arrived at the conclusion that they were not isotopes of any of the elements with atomic numbers between 85 and 92, that is of any elements near uranium.

Now in all the nuclear processes which were known up to that time unstable nuclei which were formed by neutron bombardment had atomic numbers which never differed by more than two units from the number of the bombarded nucleus. Fermi and his collaborators therefore put forward the hypothesis that a new element was involved of atomic number 93. This would be an element which occupied a new box in Mendeleev's table and which did not exist in nature. This hypothesis appeared very reasonable because it only required that uranium should behave in a manner which had already been observed for a large number of elements. For example manganese, which has atomic number $Z = 25$, when it captures a neutron gives rise to an unstable manganese which in its turn disintegrates by electron emission yielding a stable iron nucleus (of atomic number $Z = 26$).

It was thought, therefore, that in the same way uranium when it captured a neutron gave rise to an unstable isotope which decayed by electron emission and was changed into the element with atomic number 93. The fact that a new element was formed was therefore due merely to the particular circumstance that uranium is the last of the elements in the Periodic System.

The results of Fermi and his collaborators were very soon confirmed by the German chemists, Lisa Meitner and Otto Hahn, and by the Joliot-Curies.

The case of uranium however turned out to be exceedingly complex. In fact, as one tried to identify the chemical nature of the bodies which were responsible for the various activities produced by bombarding uranium with neutrons one was faced by intricate complications which were very much more difficult to unravel than they were in the case of the other elements.

The study of these phenomena required the use of extremely refined chemical techniques which were developed in particular by Hahn and his collaborators. In 1939 there appeared a report by Hahn and Strassmann of the result of a long and delicate series of chemical tests. Although almost incredulous of their amazing results they were obliged to conclude that one of the radioactive bodies (with a half life of 300 hours) which was formed by the neutron bombardment of uranium ($Z = 92$) was not, as had previously been believed, an isotope of radium ($Z = 90$) but was an isotope of barium ($Z = 56$). This discovery confronted students of nuclear physics with a type of process which was completely new and which was later given the name of *fission*—the uranium nucleus when struck by a neutron breaks into two *almost equal* parts consisting of the nuclei of two atoms of intermediate atomic number.

We owe to Frisch and Meitner the credit for having recognized, immediately after the discovery of Hahn and Strassmann, that the fission of a nucleus can easily be interpreted according to Bohr's nuclear model and that such a process could only take place in the heavier elements. Once again we will liken the nucleus of uranium to a small drop in a liquid, for example, mercury. Because of surface tension the drop tends to conserve its spherical form and only by striking it sufficiently violently can we stimulate the break-up into two smaller droplets. In this way the nucleus of uranium, which is made up of 92 protons and 146 neutrons, behaves according to Bohr's model like a drop of liquid in which 238 particles (92 plus 146) move around but maintain contact one with another. In the case of uranium the relatively small amount of energy which can be supplied to the nucleus by the capture of a neutron is sufficient to stimulate fission; for the 92 protons which it contains repel each other because of their positive charges so as to compensate a large part of the energy due to nuclear forces. A quantitative calculation shows that fission can occur as a result of the capture of a neutron of not too high energy only for the heavy elements because these, as the result of the high number of protons, are less stable.

To indicate the complication of the phenomenon of fission I need only say that it has been found experimentally that when uranium breaks up it gives rise to more than thirty-five elements distributed in the central part of the Periodic System and that by now more than 160 radioactive isotopes of these elements have been identified.

It was later recognized that uranium, when subjected to neutron bombardment, not only undergoes fission but also gives rise to the formation of trans-uranic elements of atomic number 93 (neptunium, Np), 94 (plutonium, Pu), 95 (americium, Am), 96 (curium, Cm), 97 (berkelium, Bk), 98 (californium, Cf), 99 (einsteinium, E), 100 (fermium, Fm), and 101 (mendelevium, Mv).

It is perhaps worth noting that, in addition to the trans-uranic elements, there have been produced in suitable nuclear reaction some other elements for which there were vacant

Artificial Radioactivity

places in the middle of Mendeleev's table and which do not exist in nature because they are unstable. Since 1937 there have been discovered elements of atomic number 43 (technetium, Tc), 85 (astatine, At), and 61 (promethium, Pm). As the element 87 (francium, Fr) was discovered in 1941 as a very rare branch of the actinium family there are now no new chemical elements to be discovered except possibly new trans-uranic elements.[1]

---

[1] The names of the new elements were chosen by the discoverers according to by now traditional rules: technetium because it was the first element to be artificially produced; promethium because it is produced in the fire of a nuclear reactor; astatine because, like the other halogens, it has a name which represents a characteristic peculiar to it (chlorine—green, bromine—stinking, iodine—violet, astatine—unstable); neptunium and plutonium because they are trans-uranic elements (in analogy to the name of planets); americium because it is the analogue of europium; curium (from Curie) because it is analogous to gadolinium (from Gadolin the name of a Scandinavian mineralogist); etc.

CHAPTER EIGHT

ACCELERATING MACHINES

ACCELERATION OF PARTICLES

The difficulty in exploring the intimate structure of matter consists in the sub-microscopic dimensions of fundamental particles. In the microscopic world much of the information on the structure of the objects which surround us reaches us—or let us rather say reaches our brain—by means of the eye.

But the eye is not able to distinguish objects which are smaller than about seven-thousandths of a millimetre. So we make use of optical instruments; but these, however perfect, do not allow us to distinguish details of dimensions less than about 0·5 thousandths of a millimetre, that is details about five thousand times greater than the diameter of an atom and 500 million times greater than the diameter of an atomic nucleus.

Let us try to analyse the way in which our eyes see with the aid of a microscope a very small granule. A light source emits photons which strike the granule and are scattered by it. The photons which are scattered in the direction of the microscope objective reach our eye which therefore sees the granule. This shows then that an essential phenomenon in the process which results in the vision of the granule is the 'scattering' of photons by the granule; in other words, the essential phenomenon is none other than a collision process by photons against the granule. These preliminary observations may help in orientation for what I am about to say.

In order to study the particles which make up the atomic nucleus and to study the properties of nuclear forces one always carries out collision experiments, bombarding the nuclei with very high energy particles. These particles are able to penetrate into atomic nuclei and produce phenomena which supply to

Accelerating Machines

physicists data which could not be obtained in any other way. In fact by studying the way in which a collision takes place one can determine the characteristics of the particles which have taken part in it and of the forces which have intervened.

In order to carry out a collision experiment one requires of course a machine to produce the incident particles (that is, an accelerating machine). One also requires a target against which these particles can collide; and one requires finally something which will correspond to our eyes in order 'to see' what has happened. The 'eye' which sees atomic phenomena can be of many different types such as those which I have described in the chapter on detectors.

In this chapter, on the other hand, I will explain some of the fundamental ideas on the operation of the particle accelerating machines, that is, of those machines which are used to produce the particles with which we can bombard our targets.

The only particles with sub-atomic dimensions which occur in nature are the alpha particles and the electrons emitted by natural radioactive substances and the particles of cosmic radiation (about which I shall speak later). But for many experiments natural projectiles are very inconvenient. Particles emitted by natural radioactive bodies have energies of the order of some millions of electron volts and rarely reach ten million electron volts. The particles of cosmic radiation are limited in number and mixed together in such a complicated way that experiments are often very difficult.

We can understand therefore how alongside nuclear physics there has developed the technique of the construction of accelerating machines, that is of machines which can accelerate elementary particles; and one can understand why we try to reach ever higher energies.

The particles which are accelerated in such machines can be, according to the circumstances: electrons, protons, deuterons, helium nuclei and occasionally nuclei of other light or medium elements. All of these particles have electric charge which is acted on by suitable electric or magnetic fields that are designed to produce the required increase of energy. Neutrons on the other hand are always produced by means of suitable nuclear reactions or bombardment of appropriately chosen nuclei with convenient projectiles.

The energies which the accelerating machines can now reach are of the order of some tens of thousands of millions of electron volts. At these energies the velocity of the accelerated particles differs little from that of light. It is therefore no longer possible to use classical mechanics in the design of these machines; relativistic mechanics must be used. One can also say that the satisfactory operation of the large accelerating machines now provides the most direct and clear proof of the theory of relativity which is today considered as a basic doctrine for engineering itself.

The simplest criterion for deciding whether to use relativistic mechanics or classical mechanics in the study of the movement of a particle is obtained by comparing its kinetic energy (or energy due to its motion), which we will indicate by a capital T, with its rest energy mc^2, that is the energy which corresponds to its mass when measured in the reference system in which the particle is at rest. Whenever the kinetic energy T is very much less than mc^2 one can use classical mechanics; whenever T is of the same order of magnitude as mc^2 the behaviour of the particle differs noticeably from that predicted by classical mechanics, or, as one normally says, the relativistic corrections are considerable. If, then, T is very much greater than mc^2 we are said to have an ultra-relativistic case. In fact, in this case, the particle moves with a velocity which is very close to the velocity of light and an increase in its kinetic energy does not correspond to an increase in its velocity but to an increase in its apparent mass.

In order then to know at what energy the motion of a particle should be treated by the theory of relativity it is sufficient to know the values of the rest energies of the different particles. These are given in the following table:

Values of mc^2 *in Electron Volts*

Electron	Proton	Deuteron	Helium nucleus
½ million	1,000 million	2,000 million	4,000 million

Thus an electron which has an energy of 10,000,000 electron volts is already ultra-relativistic, that is, it moves with the velocity c; whilst before a proton reaches the same condition it must arrive at an energy of a good 20,000 million electron volts.

ACCELERATING MACHINES

Having discussed these preliminary general considerations here is a brief description of the more important types of accelerators.

They can be divided into two main classes, constant voltage accelerators and variable voltage accelerators. The various types of accelerators are clearly arranged in the following table:

CONSTANT VOLTAGE ACCELERATORS: constant voltage linear accelerators.
VARIABLE VOLTAGE ACCELERATORS:
linear accelerators: linear accelerators with synchronized alternating voltage.
circular accelerators: Betatron
Cyclotron
Synchrotron
Frequency modulated cyclotron
Proton synchrotron

Before giving a summary description of the various types of accelerator I should make clear that by these expressions I mean true accelerators; but these accelerators have to be fed in the sense that one must introduce into them particles to be accelerated; in other words, an accelerator pre-supposes what is known as a source of particles. If we are dealing with electrons we require only a simple red-hot filament like that used in radio thermionic valves. If we are dealing with protons or other like nuclei the simplest source consists in a discharge tube similar to that used many years ago by Goldstein for the study of canal rays. If we want to accelerate protons we introduce into the accelerator canal rays produced in a discharge tube containing hydrogen. In the discharge the molecules of hydrogen are broken up and some of the atoms become ionized. The atoms of ionized hydrogen, that is protons, move towards the negative electrode of the discharge tube and pass through a hole bored in it to enter the accelerator.

Constant Voltage Linear Accelerators

The *constant voltage linear accelerator* was first used by Cockcroft and Walton to produce protons of above one million electron volts.

It consists of a tube of insulating material in which cylindrical electrodes are inserted, as is shown diagrammatically in Figure 31. A high vacuum is maintained inside the accelerating tube by means of a suitable system of pumps. If the tube is designed for the acceleration of positive particles (for example protons), the

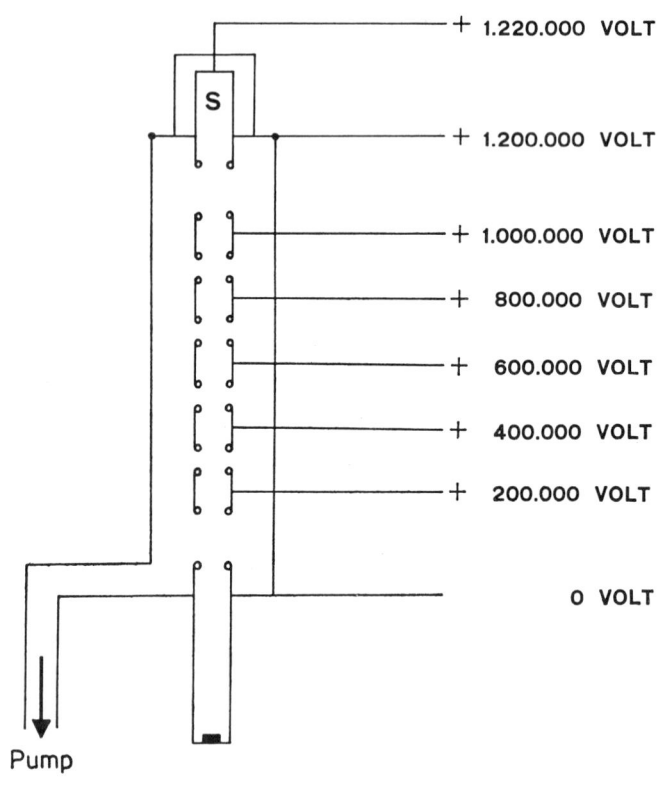

FIG. 31. Sketch of a constant voltage linear accelerator.

voltages applied to the electrodes are normally constant, increasing and positive compared to the target which represents the material to be bombarded.

The protons which are emitted by the source, S, enter the accelerating tube and under the influence of the electric field between the successive pairs of electrodes are accelerated; they reach the target with a kinetic energy which, measured in electron volts, is equal to the potential difference (in volts) applied to the extremities of the tube. In the example of Figure 31 the protons have an energy of 1,220,000 electron volts.

With machines of this type one can only reach energies of some millions of electron volts. In fact, increasing the voltage implies in general insulation of the various parts of the tube for very high potential differences and in particular as the voltage applied to the tube is increased it becomes continually more difficult to avoid discharges into the air between the external parts of the tube itself. In order to overcome this difficulty a study has been made of accelerators of the type described here which are enclosed in a huge metal envelope full of high pressure gas; but even with this device this type of accelerator cannot be used to reach energies higher than about ten million electron volts.

Variable Voltage Accelerators

In order to reach higher energies one has then to use variable voltage accelerators, in which instead of applying a constant potential difference to the particles one applies successively a number of voltage pulses.

Linear Accelerators

Machines of this type, which were conceived by Wideroe in 1929, have only been built during the last fifteen years.

The principle of operation of the simplest type of linear accelerator with synchronized alternating voltage can be very roughly described in the following way. Let us imagine that each of the electrodes in the tube of Figure 31 is connected, not to a generator which supplies a constant voltage, but to an electric oscillator similar to the oscillator of a radio transmitting station.

The protons which are emitted at a given instant may find that the potential of electrode 2 is positive compared to that of electrode 1 and will be repelled backwards; but as electrode 2 is connected to an oscillator the protons which are emitted after an interval of time which is equal to a half period of the oscillator will find that the potential of electrode 2 is negative compared to electrode 1. They will therefore be accelerated and will arrive in the region of space between electrode 2 and electrode 3. Provided that the oscillator which controls electrode

3 operates with a suitable phase the protons which arrive there will undergo another acceleration and so on. If on the other hand, the phase is wrong they will be sent back.

The metal tube of this oscillator in which a high vacuum is maintained is arranged horizontally. The electrodes are linked to the oscillators by means of conductors which pass through insulators inserted into the walls of the tube; but in this type of accelerator, unlike constant voltage accelerators, no insulator has to withstand very intense electric fields; in fact, the oscillating electric field has minimum points on the walls of the tube and reaches a maximum only in the central part of the tube.

This method is suitable for reaching very high energies because one can conveniently increase the number of sections of the tube. However, in doing so the number of oscillators must be increased and this increases the alignment difficulties and the cost of the accelerator.

The largest linear accelerator that exists today is at Stanford; it is seventy-three metres long and produces electrons of 700 million electron volts. Again at Stanford the construction of a linear accelerator three kilometres long to produce electrons up to 42 thousand million electron volts is under way.

Circular Accelerators

In all the accelerators which I shall now describe the particle to be accelerated instead of moving in a straight line moves, thanks to the presence of a magnetic field, on a circular orbit or on orbits made up of arcs of circles.

Whilst the essential characteristic of the machines which I have previously described is an accelerating tube the characteristic of the machines which we will now discuss is the presence of a large magnet whose weight sometimes reaches some thousands of tons.

Before starting on a description of the principles on which the operation of each one of these machines is based we should recall the effect of a magnetic field, H, on the movement of a particle which has charge, e, mass, m, and a velocity, v, perpendicular to the direction of the field. From elementary arguments one finds that:

(1) $$mv = (e/c) \cdot H \cdot r$$

where *e* is, as we have said, the charge of the particle, *c* the velocity of light, *H* the intensity of the magnetic field and *r* the radius of curvature of the trajectory. The product *mv* is called the momentum. As e/c is a constant the relation (1) tells us that the momentum of the particle is proportional to the product $H \cdot r$; that is, given the momentum of the particle, to bend its trajectory on to a circle of defined radius, *r*, one must choose a magnetic field, *H*, which satisfies relation (1).

Let us now start with the description of the principle of operation of the *betatron*, which is an extremely simple machine in which there is only a magnetic field. It is clearly separated from all the other circular accelerators which, as we shall see, are based on the action of two fields, one magnetic and one electric. These are closely related to each other.

The betatron conceived by Wideroe in Germany in 1929 and realized by Kerst in 1941 at the University of Illinois (United States) is a machine used for the acceleration of electrons. It consists (Figure 32) of a large electromagnet containing,

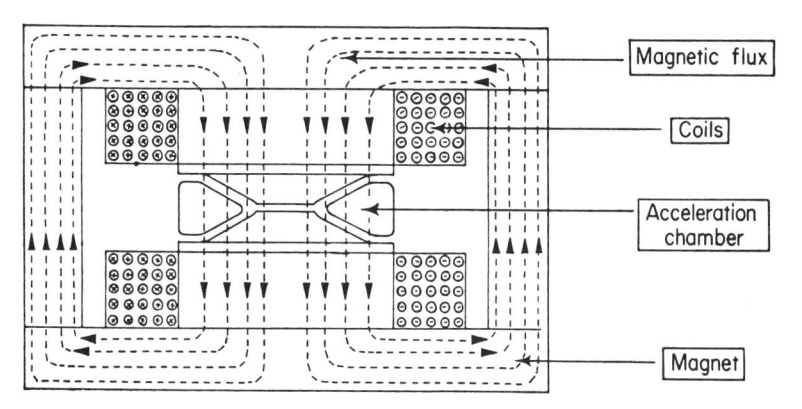

FIG. 32. Sketch of the betatron.

between suitably shaped poles, a porcelain box. A high vacuum is maintained in this box which is called because of its shape a *doughnut* (an American doughnut is in the form of a ring). The magnet is fed with alternating current and should be considered as the iron core of a static transformer rather than as an electromagnet.

Let us suppose that at a given moment in which the magnetic

field is very small some electrons are emitted by the source. Provided the magnetic field has a suitable value it will bend the trajectory of the electrons according to equation 1 in such a way that they describe, with constant velocity, a circular orbit whose centre lies on the axis of the doughnut.

The magnetic field does not only have the job of bending into a circle the trajectory of the electrons; it also accelerates them for, given that the magnet is supplied by alternating current, the magnetic field grows with time; and therefore the magnetic flux linked with the orbit of the electrons also grows. The electrons are therefore affected by a force which is tangent to their path and which accelerates them. In other words, the betatron behaves as a static transformer whose primary is made up of the windings of the magnet and whose secondary is the orbit of the electrons.

Thanks to this accelerating force the velocity of the electrons increases and will continue to increase as long as the magnetic field increases. When this field has reached its maximum value the electrons will have reached the highest energy which the betatron can produce; they are therefore ready to be used. They are directed against the target which normally consists of a thin sheet of the material which one wants to bombard. Thus each time that the magnetic field increases, that is at each period, one has a pulse of accelerated electrons which collide against the target to be bombarded. This is therefore a machine in which the acceleration and the output of the particles comes in bursts.

One can see that in the betatron the magnet performs simultaneously two jobs. By the variation of its flux it gives rise to the force which accelerates the electrons and at the same time it curves the trajectory of electrons and obliges them to remain for a relatively long time under the influence of the accelerating force.

To give an idea of the dimensions of a machine of this type it is enough to say that to produce electrons of 300 million electron volts a magnetic field of 10,000 ersted with a radius of one metre is required. The biggest betatron that exists today is that at the University of Urbana (Illinois) which supplies electrons of 350 million electron volts.

In the other machines which I shall now describe the magnetic field is used only to bend suitably the trajectory of the particles to be accelerated; an oscillating electric field is used to accelerate them. All these machines are strictly related to each other and can be considered as derivatives from the cyclotron which was invented by E. O. Lawrence in 1931-2 and built by Lawrence and Livingstone; it was then perfected by Lawrence and by his school. It is used for the acceleration of protons or of other particles of mass greater than that of protons.

The cyclotron is essentially made up of a large magnet (Figure 33a) which is fed with direct current and supplies the

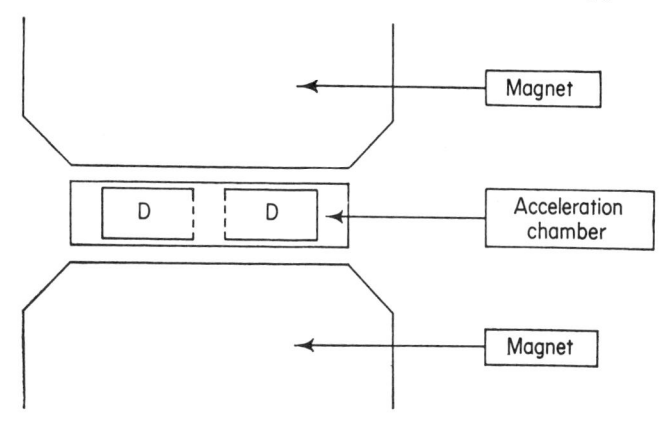

FIG. 33a. Sketch of the cyclotron.

constant magnetic field which bends the trajectory of the particles to be accelerated. Between the poles of this magnet there is a cylindrical metal box containing a high vacuum. In this box there are two electrodes which in their turn are made in the form of a thin cylindrical box which is, however, divided into two halves, one for each electrode. The two electrodes are connected to the two poles of an electric oscillator analogous to that used in radio transmitters. In this way there is an alternating electric field between the two electrodes.

If one injects, for example, a proton supplied by a suitable source near to the centre of these two electrodes it will be attracted by one of them and repelled by the other. The proton will go into the negative electrode and, as the result of the magnetic field, will describe in it a half-circle before reappearing in the space between the two electrodes. If in the time which the proton has taken to complete this half-circle the voltage

222 *The Nature of Matter*

supplied by the oscillator is inverted the proton will now be attracted to the other electrode and will undergo a new acceleration. It will enter the second electrode in which as the result of the magnetic field it will describe a new arc of a circle and so on. In this way the proton will follow a spiral path at a velocity which increases regularly until it finally collides against the target made of the material to be bombarded (Figure 33b).

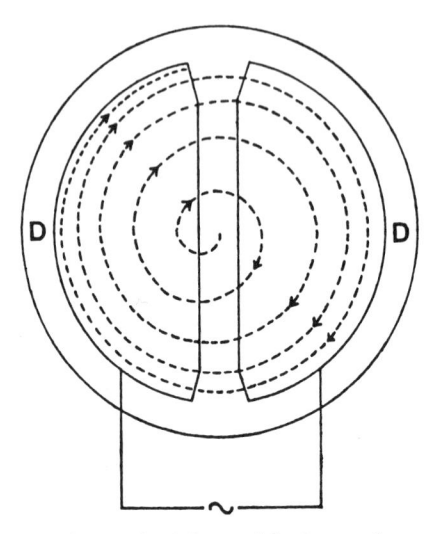

Fig. 33b. Path of the particles in a cyclotron.

If, for example, we require protons which have an energy of 10 million electron volts from a cyclotron in which the difference of potential between the two electrodes is 100,000 volts the proton must pass a hundred times from one electrode to the other, that is, it must make fifty revolutions.

One can understand therefore that the most delicate feature in the operation of a cyclotron is the synchronization condition; it is necessary that the mechanical motion of the proton should be maintained for a very large number of revolutions in phase with the oscillation of the electric potential between the two electrodes.

This synchronization condition can be deduced from equation (1) which can also be written

(2) $$v/r = (e/c)(H/m)$$

Accelerating Machines

The ratio v/r is known as the angular velocity of the particle. Now in order to maintain synchronism it is essential that the angular velocity of the particle should be equal to the angular frequency ω of the alternating current, so the condition for synchronism is

(3) $\qquad \omega = (e/c)(H/m)$

Then if H is fixed and the mass, m, of the particle is fixed one can from (3) deduce the angular frequency of the oscillator (which is equal to 2π times the number of cycles per second).

We can at once see from (3) that the cyclotron is a classical machine in the sense that for it to work it is necessary that the mass of a particle should not vary. This is the reason why the cyclotron cannot be used to accelerate electrons; for electrons require relativistic corrections which are already considerable at half a million electron volts. The cyclotron then is used to accelerate protons, deuterons, etc., and is of course limited to energies at which their masses remain constant.

If one wants to reach higher energies one must modify the machine in such a way as to conserve synchronism even when the mass of the particle increases during acceleration.

And so from the cyclotron there were born the synchrotron and the frequency modulated cyclotron.

At the end of the war (1945) Vekseler of the Academy of Sciences, USSR, and McMillan of the University of California, USA, conceived independently the idea of an accelerator for electrons which is called a *synchrotron*. In this machine, just as in the cyclotron, the acceleration of the particles is produced by an electric field whilst their trajectories are bent by the magnetic field. But in the synchrotron the particles follow a circular trajectory as in the betatron and not a spiral trajectory as in the cyclotron. In order to maintain the synchronization condition the magnetic field, H, of a synchrotron is not held constant in time but increases as the mass of the accelerated particle increases. In this way it is possible to keep constant the angular frequency, ω, of the alternating voltage providing that one injects into the accelerator ultra-relativistic electrons, that is electrons whose velocity is almost equal to the velocity of light; for, as I have already said, in this case the velocity of the particles remains constant even when their energy is increased.

The synchrotron owes its name just to the fact that in it the synchronization condition always remains valid.

In April 1959 there started to work at Frascati (Rome) a synchrotron (Plate 7a) for electrons of a thousand million electron volts. It was designed and constructed by Italian physicists, engineers and industrial firms. This is the highest energy which has been reached up to now. There are two other synchrotrons of the same energy in the United States: one at Caltec (California Institute of Technology) and the other at the Cornell University of Ithaca. The Frascati synchrotron, however, has a considerably higher intensity.

Another two synchrotrons for the production of electrons of 5 thousand million electron volts are being made, one at Cambridge (USA) and the other at Hamburg (Germany).

The *frequency modulated cyclotron* is used to accelerate heavy particles such as protons, deuterons or alpha particles. In it the magnetic field, H, is constant and the oscillator frequency is decreased as the velocity, and thence the mass of the particle, is increased. Amongst the many machines of this type those with the highest energy produce protons of 660 million electron volts—one at Dubna (Moscow, USSR) and the other at Berkeley (California, USA); the third, in order of decreasing energy is the synchro-cyclotron of Cern (European Organization for Nuclear Research, Geneva) which produces protons of some 600 million electron volts.

In order to produce protons with energies of the order of a thousand million electron volts or higher one uses *proton-synchrotrons*, that is synchrotrons for protons. The problem of acceleration of protons of these energies is more complicated than that of the acceleration of electrons for electrons are injected into synchrotrons with an energy of some millions of electron volts which is obtained for example by a linear accelerator. For this reason their velocity is already very close to that of light and therefore only increases very slightly during the whole process of acceleration. Thus the synchronism between the movement of the electron and the oscillation of the accelerating electric field can be easily obtained; but a proton on the other hand which is injected even with an energy of 50 million electron volts into a proton-synchrotron which is to accelerate it to some thousand million electron volts still

5 Photograph of a spark chamber in a magnetic field. The sparks reveal four tracks produced by secondary particles from the decays of two separate K_0 mesons (see Chapter XI); in each decay the K_0 (not seen) $\to \pi^+ + \pi^-$

6a The inside of a linear accelerator

6b The 1·5 metre constant frequency cyclotron at Argonne (USA)

Accelerating Machines

behaves according to classical mechanics. Thus, its velocity increases considerably during acceleration and synchronism can be obtained only by suitably varying the frequency of the accelerating electric field.

In other words, in the proton-synchrotron one makes use of both the principles of the synchro-cyclotron and of those of the frequency modulated cyclotron: just as in the synchrotron the magnetic field grows gradually as the energy increases in such a way as to maintain the protons on a circular orbit of a given radius; and just as in the frequency modulated cyclotron the frequency of the accelerating field varies during the acceleration in such a way as to maintain synchronism.

The two largest proton-synchrotrons in the world are in Switzerland and in the United States: that at Cern (Geneva) started to work at the end of November 1959. It produces protons of 29 thousand million electron volts; the synchrotron which produces 30 thousand million electron volts at the National Laboratories at Borokhaven (USA) started to work in August 1960. These machines are really gigantic. The Cern proton-synchrotron for example (Plate 7b) has a magnet with a radius of 100 metres. It is made up of 100 subsidiary magnets and it contains in all 3,400 tons of iron and 130 tons of aluminium. The magnet is set up in an annular tunnel which is covered with a layer of earth thick enough to prevent the radiations which are produced from harming the people who work around it. The protons are injected into the proton-synchrotron via a linear accelerator which accelerates them to 50,000 electron volts. The total time for acceleration is two seconds during which the protons travel in their circular orbit a distance which is equal to that which separates the earth from the moon.

Like betatrons and synchrotrons the proton synchrotron emits particles in bursts. Its cycle can be varied from twenty to twelve bursts a minute.

These then are the principles on which are based the huge machines which allow physicists to carry out the collision experiments which are needed in the investigation of the atomic nucleus.

CHAPTER NINE

COSMIC RAYS

THE PENETRATING RADIATION

In 1912 the Austrian physicist Hess and, independently, the Italian Pancini demonstrated the existence of a radiation which ionizes the air of the terrestrial atmosphere and whose intensity increases with height above sea level.

This radiation is not stopped in the first layers of the earth's crust but penetrates it and reaches without much loss in intensity to the bottom of the deepest lakes and into tunnels dug under the highest mountains. For this reason it was called penetrating radiation.

It was very clear from the first experiments that these rays were of extra-terrestrial origin and for that reason they were called cosmic rays. In more detail it was recognized that there exists a primary radiation which comes from inter-stellar space and invades the atmosphere of the earth; the phenomena observed by physicists are in general of a secondary nature, that is, they are due to radiations which are produced in the atmosphere by the primary radiation in its collisions against the atomic nuclei of the atmosphere.

Cosmic radiation was discovered with the same apparatus which was used for the study of the radiations emitted by radioactive substances: ionization chambers, Wilson chambers, counters, etc. It is made up of particles accompanied by gamma rays. These particles, however, have energies very much greater than those of the particles emitted by radioactive bodies.

Radioactive particles have, as we know, energies of the order of a few millions of electron volts or at most of a few ten millions of electron volts. In cosmic radiation on the other hand one frequently observes particles of some thousand millions of electron volts and sometimes of energy which is very, very much greater. The upper limit is greater than a million, million, million electron volts.

This fact shows that the study of cosmic radiation has a double interest. On the one hand it is a large-scale cosmic phenomenon (on one square centimetre at sea level about one cosmic ray arrives every minute). On the other hand, it provides opportunities for studying what happens when matter is crossed by particles with enormous energies and which can because of these energies penetrate into the depths of the most compact nuclear aggregates and in this way produce events which cannot, at least for the moment, be understood.

We should not, therefore, be surprised that physicists from all over the world have carried counters, ionization chambers, and Wilson chambers to all latitudes from Cape Horn to Greenland in order to study cosmic rays. They have sunk their instruments, equipped with appropriate systems for automatic registration to the bottom of the deepest lakes and they have carried them to the top of high mountains; they have attached them to balloons or mounted them inside a V2 or put them in artificial satellites to carry them to the highest altitudes which can be reached today. Balloons can stay at altitudes of 30 to 35 kilometres for several tens of hours and carry apparatus which weighs some tens of kilogrammes; with rockets one can reach altitudes of hundreds of kilometres for times of the order of a minute and finally artificial satellites are better than all the other vehicles in the time which is available for observation, in the distance reached and in the weight of useful apparatus which can be carried.

By these means, which are continually being perfected, one can hope to reply to the many questions which have arisen as the result of the discovery of cosmic rays: what are they? Where do they come from? How are they produced? What happens when they collide against atoms of atomic nuclei? For how long have they been invading the earth? Have they made any substantial change in it?

Some of these and other questions have received, after about fifty years of research, partial answers which can be accepted with a certain degree of confidence, but at the same time it has been discovered that the study of cosmic rays has great importance in other fields of science, which appear remote from

THE INTENSITY OF COSMIC RAYS AS A FUNCTION OF ALTITUDE

The Americans Van Allen and Tatel, using a counter mounted in the nose of a V2, measured the variation in the intensity of cosmic rays up to an altitude of 160 kilometres. The results which they obtained agreed completely for altitudes less than 35 kilometres with the results obtained in many previous experiments.

It was found that the number of particles in the cosmic radiation increases as one goes up in the atmosphere and reaches a maximum at about 20 kilometres; then between 20 and 50 kilometres the intensity of cosmic rays decreases. From 50 to 150 kilometres it remains constant (Fig. 34). At

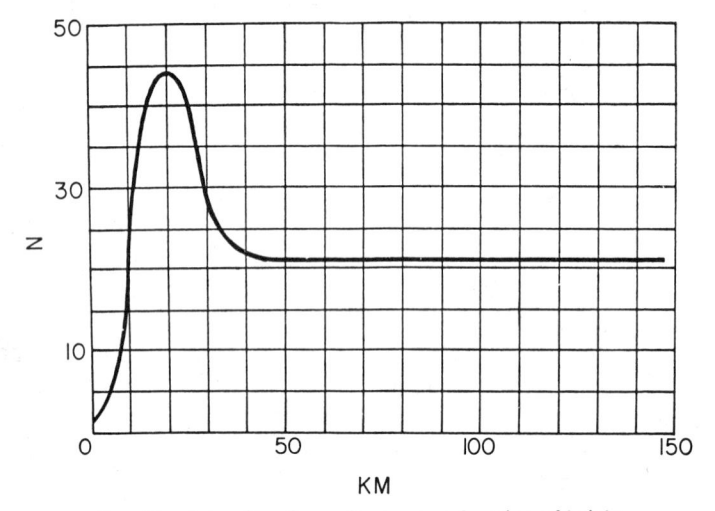

FIG. 34. Intensity of cosmic rays as a function of height.

50 kilometres one is virtually outside the atmosphere in the sense that the pressure (and thus the amount of material which is above the measuring apparatus at that altitude) has a value which is less than a thousandth of the value which it has at sea level.

The curve of Fig. 34 can be interpreted in this way. Above 50 kilometres one observes effectively only the primary radiation. The particles which make it up and which have, as we shall see, exceedingly high energies begin at that altitude to collide against the atomic nuclei of the air; in this way they produce events which I will describe later, thanks to which new particles are born at the expense of the energy of the primary particles. One has in this way an increase (between 50 and 30 kilometres above sea level) in the total number of particles. Below 20 kilometres there occurs the continuous process of absorption of all radiation (primary and secondary) by the atmosphere; and so the intensity of the radiation diminishes with height.

THE LATITUDE EFFECT

Many experiments have shown that the intensity of the primary radiation depends on the magnetic latitude (latitude effect): more specifically, the intensity of the primary radiation is lower at the magnetic equator and higher at higher magnetic latitudes.

This fact can be immediately explained if one assumes that the primary radiation is made up of particles which carry electric charge. If we consider a particle which approaches the earth from a great distance we can see that as a result of the magnetic field which surrounds the earth its trajectory will be curved (Figure 35).

The theoretical study of this problem (which is, at the same time, the study of the Northern Lights) was started by the Norwegian Störmer and carried on then by Epstein, Lemaître, Vallarta and many others. It showed that for every magnetic latitude there existed a minimum energy, below which no charged particle could reach the surface of the earth because its trajectory would be deflected backwards by the earth's magnetic field.

This minimum energy depends not only on the latitude but also on the direction from which the particle arrives. This direction has, therefore, to be specified for every case. For example, if we are dealing with protons the minimum energy in a vertical direction would be about 2 thousand million electron volts at 47 degrees of magnetic latitude (for example, near

Venice), about 4 thousand million electron volts in Sicily (37 degrees of magnetic latitude), and about 14 thousand million electron volts at the equator.

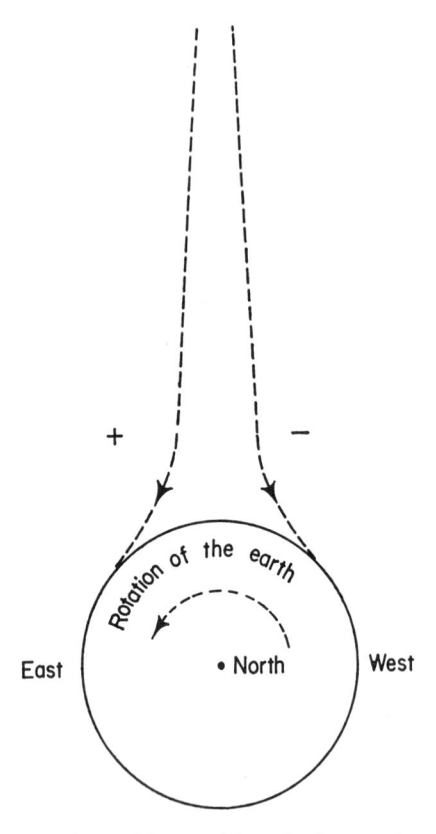

Fig. 35. Trajectories of a positive particle and of a negative particle deflected by the earth's magnetic field.

It is therefore possible to use the earth's magnetic field as if it were a huge spectrograph for the energy of the primary particles. All that is necessary is to measure the number of primary particles which fall from outside at various magnetic latitudes; from this one can deduce how many primary particles have energies greater than 2 thousand million electron volts, how many have energies greater than 4,000 Mev, how many have energy greater than 10,000 Mev. . . .

Cosmic Rays

Researches of this type have shown that the individual particles of the primary radiation have very high energy, with, as I have said, an upper limit greater than a million billion electron volts (10^{18} ev). This should be compared with the energy of the protons produced by the largest present-day accelerators—30 thousand million electron volts (3×10^{10} ev).

THE CONSTITUTION OF THE PRIMARY RADIATION

The latitude effect shows clearly that a large part of the primary radiation is made up of electrically charged particles. What type of particles are these?

Some information on the nature of the primary particles can be deduced from the following facts. As I have already said, the minimum energy with which a particle can reach the surface of the earth depends at a given latitude on the direction from which it came. If one takes two directions which make the same angle with the vertical but one towards the east and the other towards the west one finds that, if the primary particles are positive, the minimum energy will be much less for the particles which come from the east than for those which come from the west; and vice versa if the primary particles are negative. And so the measurement of this effect (called the east-west effect) should allow us to determine the sign of the charge of the primary radiation. Well, careful experiments carried out at various latitudes and various altitudes have shown without any possibility of doubt that positively charged particles predominate in the primary radiation.

It was then thought probable that the primary radiation was made up of protons; but in 1948 in one of the first flights of balloons in the stratosphere (at an altitude of 30 kilometres) it became possible to observe directly the effects of the particles of the primary radiation in photographic emulsion. Many very thick tracks were observed which crossed the whole emulsion stack. They could only have been produced by heavy particles with a high positive charge and with energies of some thousands of millions of electron volts. It was, therefore, established that the primary radiation is made up not only of protons but also of heavier particles (always positively charged).

It was discovered that these tracks were made by atomic

nuclei of the lighter elements, from the helium nucleus (atomic number $Z - 2$) to the iron nucleus ($Z - 26$). The nuclei of these elements and the protons (that is the nuclei of hydrogen) are not equally distributed in the primary radiation; the relative numbers of each correspond roughly to the relative abundances of elements in interstellar space. The curve of the abundance is a function of atomic number, has a maximum (not very pronounced) around carbon ($Z - 6$) and decreases for higher atomic numbers up till iron of which only a few nuclei are found. All these positive particles make up the primary radiation which reaches the earth from all directions at a rate which (apart from disturbances produced by solar activity[1]) is uniform in time. The intensity does not vary significantly as the earth spins about itself and revolves about the sun. This shows that in the neighbourhood of the earth cosmic radiation arrives with equal intensity from all points of the sky.

When the technique was developed and more precise measurements became possible it was found that although the cosmic radiation is made up mainly of nuclei of those elements which occur most frequently in nature nevertheless the relative numbers differ from those of the abundance of the elements in the universe (Table II). The most important difference was found to be for the nuclei of the elements lithium, beryllium and boron

TABLE II.—Chemical composition of the primary cosmic radiation

Element	Relative abundance compared to hydrogen of the more abundant elements in the primary radiation	Relative abundance compared to hydrogen of the elements in the universe (b)
Hydrogen		
Helium		
Lithium, Boron and Beryllium		
Carbon		
Nitrogen		
Oxygen		
$10 \leqslant Z \leqslant 30$		
$Z > 30$		

[1] The sun emits cosmic rays for some hours, on average about once a year, as a result of a particularly violent explosion. But these constitute a completely negligible fraction of the total observed intensity of cosmic rays.

(of atomic numbers, respectively, 3, 4, 5). These scarcely exist in the matter of the universe but they are found in cosmic radiation on its arrival on the boundaries of our atmosphere. Evidently these have been produced in the course of the voyage of the particles in space. They are produced in the break-up of heavier nuclei as a result of collisions suffered in interstellar space, where there exists on average one proton in every cubic centimetre.

For example, a nucleus of nitrogen may break up into a nuclei of lithium and of beryllium according to the reaction

$$_7N^{14} \rightarrow {}_3Li^6 + {}_4Be^8$$

Thus the cosmic radiation on its arrival at the earth carries in its make-up a trace of its past history.

What is this history?

WHERE DO COSMIC RAYS ORIGINATE?

As soon as cosmic rays had been discovered there naturally arose the question: where do they come from? The most fantastic answers were put forward: someone upheld that cosmic rays are 'the whimper of atoms being born'. Another, on the other hand, said that they are 'the death rattle of dying atoms'. But as knowledge about this radiation became more precise and detailed the problem of its origin could be confronted on bases that were more solid and scientific.

As I have mentioned in the note on page 232 it has been discovered that the sun can produce cosmic rays. In fact at the same time as the observation of certain especially intense solar eruptions an increase in the intensity of cosmic rays has been registered, an increase which is sometimes very remarkable. Clearly the sun, in conditions of special activity, emits protons (and perhaps also other particles) which have energies of some thousand millions of electron volts.

One might then think that even in normal conditions the sun continuously emits cosmic rays and that the whole of the penetrating radiation which reaches us may have an origin in the sun, but this hypothesis cannot be upheld because, as I have said, the intensity of cosmic radiation does not change from day to night.

One can then make another hypothesis. If all the other stars in our galaxies behaved as the sun—if, that is, they, at least from time to time, are sources of cosmic rays all the radiation which we observe could be the combined effect of the emission by all the thousands of millions of stars which make up our galaxy. But even this hypothesis must be rejected, for even if all the stars in the galaxy behaved like the sun one would not observe in cosmic radiation that percentage of high energy particles which is in fact observed, given that in the cosmic rays of solar origin this percentage is very small. Furthermore as the sun is very much closer to us than the other stars we should receive from this particular star a flux of cosmic rays which is very much greater; and we should then observe that periodic effect as a function of solar time which is absolutely excluded by the measurements.

One must therefore accept that if cosmic rays originate in celestial bodies in our galaxy these bodies can neither be the sun nor the other normal stars. The sources of penetrating radiation must be stars of an exceptional type.

For some time a few have accepted the hypothesis that cosmic rays originate in the so-called supernovae, that is, in those stars which exhibit the most formidable explosive phenomenon which man can witness: it is an extremely rare phenomenon. (In the memory of man only three of these have been observed in our galaxy.) A star blows up and its normal luminosity is increased by about a thousand million times. Cosmic rays could have been produced in these explosions. Investigations of the Crab nebula have supported this hypothesis. This nebula which lies in the constellation of the Bull represents the remains of a supernova explosion in the year 1054.

Today it is thought that although there may exist other important sources of primary cosmic rays which have not yet been discovered, it is very probable that at least a part of the particles in cosmic radiation originate from explosions of supernovae.

THE THEORY OF FERMI

Although the hypothesis that cosmic radiation originates only in the stars is very plausible it has not been securely demonstrated. In the meantime a few physicists have considered an alternative possible mechanism which was first suggested by

Fermi in 1949 and which has certain extremely interesting facets: the particles which make up the primary radiation would according to this mechanism have been emitted by stars with a relatively low energy and would then have been accelerated during their journey through interstellar space by the effect of the immanent electromagnetic field.

Interstellar space is not completely empty; there exists in it that which is called *interstellar matter* which is made up on the average of about one atom of hydrogen per cubic centimetre. This then is matter of exceedingly low density; one need only remember that inside a cyclotron or the tube of an accelerator in which there is an exceedingly high vacuum there remain about a thousand million molecules per cubic centimetre.

This matter which fills the interstellar space is not however uniformly distributed but is condensed into *clouds*; these clouds, of dimensions of the order of hundreds of billions of kilometres, travel with a velocity of about thirty kilometres an hour; the space between one cloud and another is very much more rarefied.

The conditions in interstellar space are such that all of the atoms are ionized. This arises from the very low density of interstellar matter, both in the cloud and in the space between the clouds and because of the very high intensity of light radiation which is emitted from the stars and which floods throughout space in all directions producing from time to time photoelectric effects.

As a result of these conditions interstellar space is an excellent conductor of electricity; and the movements of these enormous clouds are accompanied by movements of electric charges which generate in the surrounding space magnetic fields; these magnetic fields vary from point to point and from moment to moment in an extremely complex and chaotic manner.

Thus interstellar space which we are accustomed to think of as empty should instead be thought of as full of enormous and extremely rarefied clouds which travel in all directions carrying with them their magnetic fields; those magnetic fields, although very weak, extend over enormous regions of space.

And according to Fermi this is how a primary proton in the cosmic radiation can acquire its high energy. Let us suppose that a proton, that is an atom of ionized hydrogen, is travelling through space; every so often it encounters a cloud whose magnetic field will deflect its path. Such an encounter can be described as a type of collision between the proton and the cloud; in this collision the proton may gain or may lose energy. Fermi has shown that the dominant collisions are those in which the proton gains energy and that providing one waits for long enough the protons will tend to reach an energy equal to that of the cloud; the kinetic energy of the cloud is enormous and corresponds to about 10^{50} electron volts.

There exists then in interstellar space a mechanism capable of accelerating protons. But from time to time one of these may undergo a collision with another proton, one of those which are stationary in space. A collision of this sort gives rise, as we shall see, to a shower of particles: the result is a loss of energy by the incident proton and the production of other particles including a fast proton which in its turn can be accelerated by means of collisions with the clouds.

We have seen that protons can undergo two types of process: collisions with clouds of dimensions of the order of 10^{19} centimetres in which on average they become accelerated; and collisions with other protons in which, on the one hand, they lose energy and, on the other hand, generate other protons which are fast enough to start again the same story of acceleration by collision with the clouds.

This theory has the merit of endowing the primary particles of cosmic radiation with an energy spectrum of the same type as that found experimentally. But according to this theory the increase in energy which a proton obtains on average in every collision is extremely small, of the order of a few tens of electron volts. Therefore it is the oldest protons which have the greater energy; but as we shall see it has been discovered that the particles which reach us have an age of about a million years; there should not therefore exist in the penetrating primary radiation particles with an energy greater than that which they can acquire in this period of time by collision with the clouds of interstellar space; they should not have an energy greater than 10^{12} electron volts. This conclusion is not confirmed by the results of observations: there are cosmic ray particles which have energies greater than 10^{18} electron volts.

Various improvements of this theory have been suggested and these would perhaps allow this difficulty to be overcome. But only further research and measurements can decide if this acceleration mechanism for cosmic ray particles can constitute if not the main process, at least a secondary process which contributes to the production of the cosmic rays which we observe.

THE JOURNEY OF COSMIC RAYS THROUGH SPACE

If, as appears reasonable, the cosmic rays which we observe on earth originate in our own galaxy they cannot reach us directly as, for example, occurs for light emitted by the stars. In fact the stars of this battered spiral system which is the galaxy are not distributed symmetrically compared to the earth; it, with the solar system, is situated in an eccentric position (to be precise in one of the arms of the spiral); thus if the cosmic rays travel along straight lines most of them would reach us from that part of the galaxy in which is collected the greater number of stars, that is, from the Milky Way. But this does not happen. The fact that the particles of the primary cosmic radiation are isotropically distributed shows that during their journey through interstellar space they have been scattered through large angles in such a way that when they arrive they have completely lost the 'memory' of the original direction in which they were emitted. These deviations could only be produced by the extensive and irregular interstellar magnetic fields: this is the only plausible mechanism which can account for the isotropy of the primary particles which arrive in our atmosphere.

Cosmic rays then follow a tortuous path through space and are imprisoned in the galaxy for a time which, as I have said, cannot be greater than a few million years. In fact the radiation which reaches us contains a certain number of fairly heavy atomic nuclei; these cannot have encountered in interstellar space a large amount of material; had they done so all these heavier nuclei would have been broken up by collisions into their constituent particles, protons and neutrons; these neutrons being unstable would be transformed into protons. Thus the cosmic radiation would finally be composed only of protons.

Knowledge of the composition of the primary radiation and of the density of matter in interstellar space allows us therefore to establish an upper limit to the time for which cosmic rays can be imprisoned in the galaxy: and this limit is a few millions of years.

COSMIC RADIATION IN THE ATMOSPHERE

When the extremely high energy particles of the primary radiation enter the atmosphere, they collide with the atomic nuclei in it and give rise to a succession of interlaced nuclear interactions; the result of these interactions is the complex, secondary cosmic radiation which arrives at sea level.

The catastrophic interactions of the primary particles with the nuclei of the first layers of the atmosphere already produce secondary particles and these too are in general of very high energy. Some of these with very high penetrating power succeed in crossing the whole of the atmosphere and arriving uninterrupted at sea level; others on the other hand interact in their turn with the nuclei of the atmosphere and give rise to new radiation; and so on. There are gamma quanta which give rise to those effects which result in the ejection of electrons or the production of both positive and negative electrons; these are often very fast. There are positive and negative electrons with extremely high energy which, when they pass through matter, not only ionize the atoms but also irradiate gamma quanta, just as in an X-ray tube the electrons incident on the anticathodes emit this radiation as the result of the abrupt deceleration which they undergo. There are protons and neutrons, the products of the numerous disintegrations undergone by the nuclei in the air and these in their turn can produce further disintegration.

Not all of these particles are stable. The electrons, both positive and negative, and the protons are stable; but one cannot say the same of the neutrons which have a mean life of about twelve minutes and which disintegrate into protons and electrons.

THE MESON IN THE COSMIC RADIATION

As I have already said the positive electron was discovered in cosmic radiation by Anderson and by Blackett and Occhialini in 1932. In the years which followed there was discovered in the cosmic radiation another particle to be added to the list of those already named: neutrons, protons, electrons and positrons (to which there was to be added in 1934 the neutrino).

In 1933 Kunze, a German, examining an extremely large number of cosmic particles photographed with the Wilson chamber noted that one of the tracks showed an ionization which was too large to be due to an electron and too small to be due to a proton. Other photographs of tracks, although extremely rare, appeared very strange. Various interpretations were put forward until finally in 1936 the American physicists Anderson and Neddermayer became convinced of the existence of a new particle which should have a mass intermediate between that of the electron and that of the proton.

There were very many beautiful and very important experiments carried out in order to establish the identity of this new particle: these experiments showed that it is extremely penetrating and has an electric charge which may be either positive or negative and whose value is equal to the charge of the electron (that is, to the unit charge); the experiments also showed that the particle disintegrates with a mean life of about two microseconds and has a mass which is equal to about two hundred times the mass of the electron: that is, of the same order of magnitude as that of the particle which had been conceived by Yukawa in order to explain nuclear forces, i.e. *the meson*.

It is easy to understand that such a coincidence—at least in order of magnitude—between the mass of the new particle and the mass of Yukawa's meson immediately gave rise to very great interest. Was this coincidence just due to chance or were the new particles which reach us from the atmosphere effectively just those which, in the way proposed by Yukawa, determined the attractive forces which link together the heavy particles which make up nuclei? This idea appeared to be supported by a particular property of the new particle. In 1935 Bernardini and Bocciarelli noted that for a given amount of matter the more rarefied the air of the atmosphere, the more effectively it absorbed the cosmic radiation. This strange phenomenon was definitely established in 1937: for a given amount of matter traversed the new particles undergo greater absorption in rarefied matter (for example air) than in dense matter (carbon, lead, etc.).

The interpretation of this fact, put forward in 1938, is clear if one accepts that these particles are unstable and can therefore disintegrate along their path. For they would require a much longer time to cross for example two kilometres of air than they would need to cross a layer of lead containing the same amount of matter (about ten centimetres). In the layer of air there would be time for the disintegration of a greater number of particles and therefore it would appear to us that the layer of air had a greater absorbing power than that of the lead. We can understand therefore how, by comparing the absorption which the particles undergo in a layer of some kilometres (for example between the peak of a mountain and sea level) and the absorption which they undergo in a layer of lead containing the same amount of matter, it is possible to deduce their mean life.

It was found from a very large number of experiments that the mean life of this particle, that is the interval of time for which, on the average, it can exist before disintegration is 2·20 microseconds.

What happens to a meson which dies? According to the idea of Yukawa it should disintegrate into an electron and neutrino. Well, measurements carried out by Anderson in a Wilson chamber in 1948 and 1949 showed that the spectrum of the disintegration electrons from mesons is a continuous spectrum, in some ways analogous to that of normal radioactive nuclei. From the fact that this spectrum is continuous it follows, on the basis of arguments similar to those used to interpret beta radioactivity, that the emission of electrons must necessarily be accompanied by the emission of two neutrinos; that is, the disintegration process of mesons is

$$\text{meson}^+ \to e^+ + 2\nu$$
$$\text{meson}^- \to e^- + 2\nu$$

At this point we might feel justified in concluding that the new particles discovered in cosmic radiation are just those mesons introduced by Yukawa in order to explain nuclear forces; they have in fact a mass of the right order of magnitude and they disintegrate, emitting electrons.

But experiment showed that the two mesons—the theoretical

Cosmic Rays

one postulated by Yukawa and the one observed at sea level in cosmic radiation—are profoundly different. Whilst the meson of Yukawa's theory—in accordance with the principle of the reversibility of microscopic processes—should be emitted and absorbed with the same probability the particle observed in cosmic radiation behaves in a distinctly different manner.

This behaviour was studied by the Italians, Conversi, Pancini and Piccioni; they set out to investigate the behaviour of mesons from cosmic radiation once they had been stopped in a piece of matter. Positive mesons once brought to rest wander through matter with a velocity which corresponds to thermal agitation until they decay according to the process indicated above. Negative mesons, on the other hand, once reduced to rest are attracted by the nuclei of the surrounding matter and therefore become 'captured' in orbits similar to that travelled by the electron. In other words, negative mesons, start to revolve about a nucleus, first in a large orbit, which implies a small binding energy, and then jumping into a smaller orbit emitting energy in the form of photons; then they jump to a still smaller orbit, and so on until one of the two following things happens: either the meson disintegrates according to the process indicated above or it is absorbed into the nucleus. Now Conversi, Pancini and Piccioni not only discovered in 1946 the existence of these strange atoms called 'mesic atoms' in which a meson instead of an electron revolves around the nucleus; they also discovered an unexpected phenomenon (which is called after them): in the nuclei with high atomic number, Z (such as, for example, iron $Z = 26$) the nucleus absorbs the negative mesons so quickly that they have no time to disintegrate; on the other hand in the light elements such as carbon ($Z = 6$) the nucleus has so little 'appetite' for mesons that in most cases they decay before being absorbed.

The full importance of the Conversi–Pancini–Piccioni effect was made clear shortly after its discovery by Fermi, Teller and Weisskopf who showed that such behaviour was incompatible with the properties which Yukawa had foreseen for his meson; this for any value of the atomic number should necessarily be absorbed so quickly that there would be no time for them to decay spontaneously. The absorption of negative cosmic ray mesons by light nuclei was about a billion times too small compared with that predicted from their production in the upper atmosphere by the primary cosmic radiation.

In other words, the mesons of cosmic radiation have qualitatively but not quantitatively the properties of Yukawa's meson and the discrepancy as we have said is enormous.

So the meson of cosmic rays is not the meson of Yukawa?

THE μ MESON AND THE π MESON

This question was answered in 1947.

The American physicist Marshak suggested in that year that the strange discrepancy between the theoretical meson and the cosmic ray meson could be resolved if one accepted the existence of *two* different types of meson, both with masses between the mass of the electron and that of the proton: one, the heavier, is responsible for the powerful nuclear forces and has the properties of Yukawa's meson (amongst others the probability of being emitted and absorbed by nuclei); the other, the lighter, is that which is observed in cosmic radiation at sea level and is produced in the decay of the heavier meson; it should have a mean life about a hundred times shorter than the meson studied by Conversi, Pancini and Piccioni.

When the primary cosmic radiation, coming from interstellar space, penetrates into the atmosphere some of the constituent particles interact with the nuclei of the atmosphere and give rise to the production of mesons of the heavier type; these start their journey through the atmosphere towards the earth; but it is a very short journey because on the average after a hundredth, millionth of a second (the value of their mean life) each of them decays, giving rise to a meson of the lighter type; this, which has the property of being only with difficulty absorbed by a nucleus, succeeds in crossing the atmosphere and reaching sea level where it is observed in our apparatus.

This was the only satisfactory way of accounting for the strange properties of a particle which in the uppermost layers of the atmosphere showed itself to be easily emitted by nuclei, yet which was so reluctant to be absorbed by them that it was able to cross the whole atmosphere and reach sea level undisturbed.

After this theory had been suggested only a few weeks elapsed before Lattes, Occhialini and Powell discovered in a photographic emulsion exposed at an altitude of 3,400 metres above sea level (in the Bolivian Andes) the track of a meson which was heavier than that which had been observed at sea level and which decayed just in the manner previously suggested. Many other subsequent experiments made not only with cosmic rays but also in the laboratory using powerful accelerating machines have since shown that this heavier meson, in contrast with the lighter one, interacts strongly with protons and neutrons. It has just the property of that meson which had been introduced theoretically by Yukawa.

This heavier meson was given the name of *pi meson* (π) or more briefly *pion*; this is called the *mu meson* (μ) or *muon*. The pi meson is Yukawa's meson, the quantum of the nuclear field, or the cement which binds together the constituents of the nucleus.

'It is really amazing,' wrote Leprince Ringuel, 'that the concept of mesons should have been forced progressively on theoretical physicists in order to explain the interaction of the nuclear field and that the mass and the mean life of this particle should have been announced by Yukawa at least to an order of magnitude in his first publication; and that on the experimental level the discovery of the meson was obtained as a result of long efforts and important experiments without any relation to the theoretical research on nuclei.'

We now know that the pi meson has a mass which is about 270 times greater than the mass of the electron, whilst the mass of the mu meson is about 205 times greater than that of the electron. Both types of meson decay spontaneously: if left alone in empty space a pion decays in a few hundredth, millionths of a second (about $2 \cdot 5 \times 10^{-8}$ seconds) into a muon and a neutrino. Depending on whether the pi meson is positive or negative the disintegration occurs according to the processes (in which I have indicated as is usual the neutrino with the Greek letter ν)

$$\pi^+ \to \mu^+ + \nu$$
$$\pi^- \to \mu^- + \nu$$

Afterwards the mu meson, positive or negative, decays in about two millionths of a second into an electron (positive or negative) and two neutrinos:

$$\mu^+ \to e^+ + 2\nu$$
$$\mu^- \to e^- + 2\nu$$

There exists also the neutral pi meson which decays into two gamma rays in an extremely short time (about one thousandth billionth of a second).

And so to that list of elementary particles which already appears to be too long (photon, electron, proton, neutron, positron, neutrino) the study of cosmic radiation has added two further particles (the muon and the pion). But this list is still far from being complete: we shall see this in the last chapter.

Let us return again for a short time to cosmic rays and see what are the phenomena which develop in complicated succession during their voyage through our atmosphere.

COSMIC RAY PHENONEMA IN THE ATMOSPHERE

The most important phenomena which cosmic radiation produces as it moves through the atmosphere are:

Showers
Stars
Penetrating showers
Mixed showers
Extensive showers

The *showers* were discovered by B. Rossi in 1931. Let us imagine that an electron with an energy of some hundreds of millions of electron volts crosses matter. It will not only ionize the molecules, it will also radiate high energy gamma quanta, which will in their turn produce electrons by the Compton effect and by the photo-electric effect and above all in pairs of high energy positrons and electrons. These in their turn will irradiate gamma quanta and so on; we have the formation of that which is called a *shower*. The process of the formation of showers which I have just described is normally called a *cascade process*.

By *star* we mean the disintegration of an atomic nucleus from which are emitted one or more heavy particles: these may be nucleons, alpha particles, or light nuclei (as for example lithium nuclei or beryllium nuclei). The name of *star* was given to this phenomenon because of the form in which it presents itself in a Wilson chamber or in a photographic emulsion (plate 4a). This typically nuclear phenomenon, which was discovered by Blau and Wambacher in 1937, may be due to a pi meson but is more often produced by a proton or a neutron of 100–200 million electron volts; these are relatively frequent at some thousands of metres above sea level.

The *penetrating showers* discovered by Wataghin and by Janossy in 1940 are due to extremely high energy protons or neutrons (of a thousand million electron volts or greater). A nucleon of this energy when it traverses a nucleus collides with another nucleon and produces at the expense of its own energy a certain number of pi mesons. From the struck nucleus there are also emitted some fragments similar to those emitted in a star. We have here a process of production of mesons by the collision of one nucleon with another; this process is analogous to the production of these particles obtained with a cyclotron. The only difference is that in the case of penetrating showers a considerable number of particles is emitted and they have extremely high energy.

Mixed showers which were observed in 1947 by Rossi and his collaborators are similar to penetrating showers; but in the mixed showers we have also the production of positive and negative electrons which are generated by the gamma rays from the decay of π^0 mesons; these are produced together with the mesons π^+ and π^- in the collision between two nucleons.

Finally there are *extensive showers* which were discovered by Auger in 1938, and which are similar to mixed showers; they differ from them only in the scale of the phenomenon. They are produced in the upper atmosphere by an extremely high energy primary (10^{15}–10^{16} and more electron volts), which collides with a nucleon of a nucleus in the air. They then yield many positive, negative and neutral mesons (in addition to other particles of which I shall speak in the last chapter); these mesons are so energetic that they can produce in their turn other particles and so on. The extensive shower crosses the whole atmosphere and when it reaches a mountain altitude or sea level is made up of millions, sometimes thousands of millions of particles which fall together like rain over an area of many acres.

From this brief description of the more important phenomena which are observed in cosmic radiation one can clearly understand the interest which is aroused by these investigations.

The showers in which only electrons and gamma quanta are involved are phenomena of an electro-dynamic nature; the stars are typically nuclear processes; the penetrating showers are phenomena in which is exhibited in a direct manner the nature of the interaction between nucleons of very great energy; extensive showers and mixed showers are phenomena which indicate, among other things, the production of photons by electrically neutral pi mesons.

Let us now follow what happens from the moment in which a proton of the primary radiation penetrates into the atmosphere.

It very quickly undergoes a first collision with an atomic nucleus in the very high atmosphere and in this collision sets in motion some nucleons and produces a shower of pi mesons. Continuing its journey through the atmosphere, this primary proton sooner or later undergoes a second, third and perhaps even a fourth collision similar to the first. Most of the secondary particles which are produced in these collisions have themselves energies high enough to produce in collisions with other atomic nuclei in the atmosphere recoil nucleons and showers of pi mesons.

By means of these processes the primary radiation is absorbed so that it has practically disappeared after crossing about a third of the atmosphere. But, on the other hand, the particles produced in these collisions—recoiling nucleons, pi mesons and also electrons generated by the gamma quanta from π^0-decay—increase in number with increasing depth in the atmosphere; their intensity reaches a maximum (which is very clear in Fig. 34) after which their number decreases since now the absorption of the secondaries prevails over their production by the primary particles.

In the middle and lower atmosphere the secondary nucleons of the cosmic radiation have relatively low energies and thus there is a decrease in the number of pions which they produce with a corresponding increase in the number of star type nuclear disintegration. Few pions arrive at these levels because they decay higher up (as the result of their short mean life) and give rise to mu mesons and to neutrinos.

In this way we can understand how the composition of the cosmic radiation changes noticeably with height above sea level: outside the atmosphere almost only protons, at very high altitudes protons, neutrons, pions and a few electrons; then, as they reach the ground always fewer high energy protons and neutrons, always fewer pions: and in the lower atmosphere mu mesons, positive and negative electrons and relatively low energy neutrons prevail.

The history of a primary, medium or heavy, element which falls on the earth's atmosphere is not substantially different from that of a primary proton; the essential difference lies in the fact that often compound nuclei are broken up in the first collision into nuclei of lighter elements or directly into all of their component nucleons.

Everything which I have said above displays the interest attached to the study of cosmic rays, to the study of these particles which reach our planet 'like a regular hail' writes Leprince Ringuet, 'continuous, implacably constant; an imperturbable hail which pays no attention to the time of day, the season of the year or the position of the sun or of the moon or of the Milky Way; a hail which surrounds everything, which passes through everything with a frequency of some millions of particles a day.'

The study of this radiation, which was unknown fifty years ago, has made possible the observation of phenomena which (because of the enormous energies in play) would not otherwise be observable and to obtain in this way fundamental information on the structure of the nucleus.

But not only this. Astronomy also has acquired a new and extremely important tool for research. Classical astronomy was based exclusively on the observation of the light radiation which reaches us from the various celestial bodies. Today to this 'optical astronomy' there has been added two new branches of astronomy: radio astronomy, which supplies information on the celestial bodies from the examination of the radio waves which they emit, and the study of the primary cosmic radiation. This radiation, thanks to the electric charge carried by its particles, brings to us a message, often even today indecipherable, about the electric and magnetic fields which exist in the celestial bodies and in interstellar space.

CHAPTER TEN

NUCLEAR ENERGY

THE BINDING ENERGY OF NUCLEONS

Let us return to and consider in greater detail the arguments, developed in the chapter on the nucleus, concerning the binding energy of nuclei in order to show how, in suitable conditions, it may be possible to use this energy for practical purposes.

I have already said that the nucleons which make up a nucleus are strongly bound together and that the mass which apparently disappears when they are bound to each other corresponds to the *binding energy* of the nucleus. The simplest example is supplied by the simplest element having a nucleus which is made of more than one particle, that is of heavy hydrogen (deuterium), whose nucleus (the deuteron) is formed of a proton and a neutron. As the atomic weight of hydrogen is equal to 1·0081 units of atomic weight and as the atomic weight of the neutron is equal in the same units to 1·0090 the total weight of the particles making up the atom of deuterium is

$$1·0081 + 1·0090 = 2·0171$$

From direct measurements on the other hand it is found that the atomic weight of deuterium is equal to 2·0147. Thus the mass defect of the deuteron is equal to

$$2·0171 - 2·0147 = 0·0024$$

units of atomic weight. According to Einstein's relation ($E = mc^2$) there corresponds to this mass defect a binding energy equal to

$$0·0024 \times 931 = 2·23 \text{ million electron volts}$$

Thus the proton and the neutron which make up a nucleus of deuterium are bound together with an energy which is about

Nuclear Energy

250,000 times greater than the energy which holds bound together the two atoms of the molecule of hydrogen.

In absolute agreement with the results of this calculation experiment has shown that when a proton captures a neutron to form a deuteron there is emitted (in the form of gamma radiation) an energy of 2·23 million electron volts. This perfect agreement between the theoretical predictions and the experimental results has now been reconfirmed in an enormous number of measurements; it establishes not only the validity of Einstein's equivalence between mass and energy but also the theoretical interpretation of the mass defect as the binding energy of nuclei.

In the chapter on the nucleus we calculated the binding energy of the nucleus of helium (that is of the alpha particle) and we found that it is equal to 28,000,000 electron volts; that is each of the four nucleons which make up an alpha particle is bound on average with an energy of $28/4 = 7,000,000$ electron volts. A similar calculation for the element next to helium, that is lithium (atomic number 3) shows that the binding energy of its nucleus is 30·5 million electron volts; that is each of the six nucleons (three protons plus three neutrons) in its nucleus have on average a binding energy equal to about 5,000,000 electron volts.

If we now compare the binding energy of the nucleons in the deuteron, in the alpha particle and in the nucleus of lithium (respectively about 2·2 Mev, 7 Mev and 5 Mev) we note that in the alpha particle the nucleons are bound together with an energy which is greater than that of the nucleons either in deuterium or the nucleus of lithium. This indicates that the alpha particle is a particularly stable system; experiment fully confirms this conclusion.

The binding energy, E, of a nucleus divided by the number, A, of nucleons which make it up represent then on the average the energy with which each of the nucleons is bound inside that nucleus. This ratio, $\epsilon = E/A$, is given the name of mean binding energy per nucleon.

By measurements of the mass of the isotopes of the various elements it has been found how the binding energy per nucleon varies with the atomic weight, that is with the variation in the number of the particles which make up the nucleus. The curve which is obtained is shown in Fig. 36. We can see that the average binding energy per nucleon is not constant but varies as one moves up from the deuteron to elements of greater atomic number.

Evidently (as I have already noted in the case of the alpha particle, that is the nucleus of helium) the greater the mean binding energy of its constituent parts the greater the stability of a nucleus. The diagrams exhibit therefore a very important

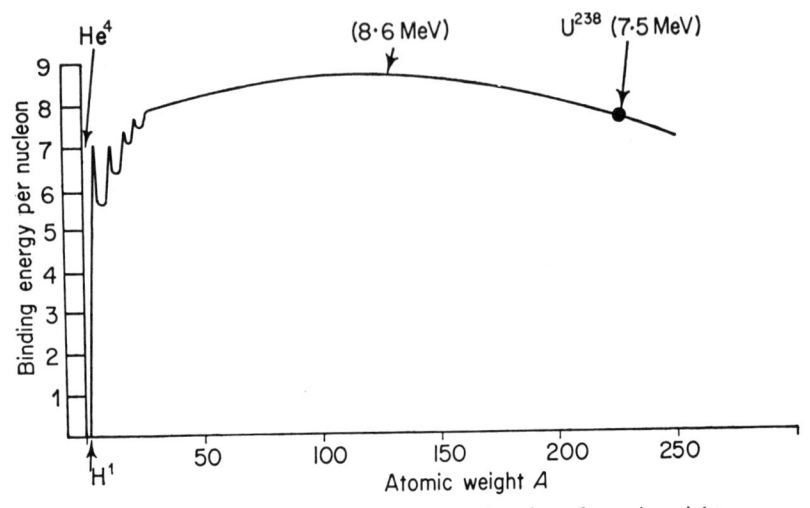

Fig. 36. Binding energy per nucleon as a function of atomic weight.

result: that the more stable nuclei are those of intermediate atomic weight. In fact if one ignores the details of the curve it shows by its general development that the mean binding energy of the nucleon starts from small values from the light nuclei (2·2 Mev for the deuteron) increases until it reaches a maximum (about 8·5 Mev) for nuclei with intermediate atomic weight, and then diminishes for the heavier nuclei.

The general development of this curve can easily be explained.

Let us consider first the left side of the curve and see why nuclei containing few particles have a smaller mean binding energy per nucleon. The binding energy of every particle which is part of a nucleus originates in its interaction with the other nucleons which are in its immediate vicinity. Thus a nucleon which is surrounded on all sides by other nucleons makes a bigger contribution to the mean binding energy than that of the nucleon which is at the surface of the nucleus and which therefore interacts with other nucleons only on one side. Now the total binding energy of the nucleus is made up of contributions from all these constituent nucleons both internal and peripheral; the latter however make a smaller contribution. But the nucleus which contains a small number of particles has relatively a larger number of peripheral nucleons than a nucleus made up of many particles.[1] The contribution made by the peripheral nucleons is the more important the smaller the number of nucleons, just as in a drop of liquid whose surface energy per unit mass is very much greater when the drop is very small. Consequently, as our diagram shows, the mean binding energy per nucleon will be smaller for nuclei made up of few particles.

Let us now consider the right-hand side of the curve. The gradual descent of this curve at the extreme right is essentially due to the electrical forces of repulsion. In fact as one increases the number of neutrons and protons which make up the nucleus (that is, as one considers the heavier nuclei) the electrostatic forces of repulsion which act between the protons also increase; and although this effect is only small compared to the nuclear forces it nevertheless becomes more and more significant and leads in the very heavy nuclei to a reduction in the mean binding energy for each nucleon.

The particular shape of this curve which represents the binding energy per nucleon as a function of the number of nucleons leads to an exceedingly interesting result.

EXOENERGETIC AND ENDOENERGETIC REACTIONS

Any reaction whatsoever, whether it be a chemical reaction or a nuclear reaction, takes place either with the production or with the absorption of energy. Reactions of the first type are called exoenergetic and of the second type endoenergetic reactions. For example, the formation of water (H_2O) from hydrogen and oxygen is an exoenergetic reaction which takes place with the production of energy (thermal energy); whilst the

---

[1] The constituent particles of nuclei made up of two, three or four nucleons can all be regarded as at the surface.

formation of acetylene (C_2H_2) starting from the elements C and H is an endoenergetic reaction which takes place therefore with the absorption of energy. A compound which is formed by means of an endoenergetic process, that is with the absorption of heat yields when it breaks up to an exoenergetic process, that is with the production of heat; in particular the break-up of compounds which are formed in strongly endothermic processes give rise to very large amounts of energy: they are explosive. For example, acetylene explodes even on its own, especially when it is liquefied or compressed, breaking itself up into the elements which make it up (carbon and hydrogen).

In the field of nuclear reactions the transformation for example of lithium into helium by bombardment with a proton and the emission of an alpha particle, that is the reaction

$$_3Li^7 + p \rightarrow {}_2He^4 + \alpha$$

is an exoenergetic reaction. On the other hand the following reactions are endoenergetic: (n, p), (γ, p), (γ, n), (p, n). For example a nucleus of carbon of atomic number 6 and atomic weight 12, when struck by a gamma quantum, is changed into a nucleus of carbon with atomic weight 11 (i.e. an isotope of the previous nucleus) and emits a neutron: that is,

$$_6C^{12} + \gamma \rightarrow {}_6C^{11} + n$$

In order that this reaction should take place the incident gamma quantum must have an energy of at least 20,000,000 electron volts.

From a scientific point of view an endoenergetic reaction has the same importance and the same interest as an exoenergetic reaction. But from the point of view of practical application only exoenergetic interactions are important. It is only to these that we must turn if we seek to make use of the enormous energies contained in atomic nuclei.

FUSION AND FISSION

As a general rule we should therefore expect that a nuclear interaction should be exoenergetic when its effect is a change from nuclei having smaller binding energy to nuclei with greater binding energy since in these processes the total mass

Nuclear Energy

is decreased; this decrease multiplied by c^2 is just the liberated energy.

Let us turn then to Figure 36 which shows that the particles which make up the nuclei of elements with intermediate atomic weight are bound together with an energy which is greater than that of the nucleon of the elements both with lower atomic weight and with higher atomic weight. This means that there are two types of exoenergetic reactions: there are those which start from light nuclei and lead by *fusion* to heavier nuclei; and there are those which start from heavy nuclei and lead by *fission* to lighter nuclei. These two types of reaction result in the production of energy: fusion and fission; they are represented in the figure by the two shifts from the two boundaries of the curve towards its centre.

There are then two ways of obtaining the production of nuclear energy: the first is that of the *fusion reaction* of the lighter nuclei; the second is that of the *fission reaction*.[1]

The nuclei which certainly cannot be used to obtain exoenergetic interactions are those of medium atomic weight, in the highest part of the curve. In fact starting from these nuclei both of the reactions of fusion and the reactions of fission are exoenergetic reactions (that is, reactions which take place with the absorption of energy), since they lead from nuclei which are more tightly bound to nuclei which are less tightly bound.

We have therefore two processes for obtaining nuclear energy: either we start from very light nuclei in order to build up nuclei which are less light, or we break apart very heavy nuclei. Crudely one can say that these two processes have analogues in the field of chemistry; indeed when molecules of oxygen are mixed with molecules of hydrogen to form water ($2 H_2 + O_2 = 2H_2O$) an exoenergetic interaction takes place in which a considerable quantity of heat is produced: in fact the temperature of the ensuing flame reaches 2,800 degrees centigrade. This in the chemical field is a reaction analogous to the fusion reaction in nuclear fields. Mercury fulminate on the other hand—a mercury compound whose formula is $Hg(CNO)_2$ —supplies in the chemical field an example of an exoenergetic

---

[1] The term fusion is used in the sense of 'fusing', or melting together; the word fission has the same root as fissure; it is also applied in biology to describe the division of a cell into two new cells.

interaction analogous to fission; indeed when struck it explodes, that is breaks up suddenly into its component parts giving rise to a very sudden gas expansion at high temperature and an explosive wave which is propagated with a speed of about 3,000 metres per second.

FUSION OR THERMO-NUCLEAR REACTIONS

As I have already said in the fusion reactions lighter nuclei fuse together in order to make heavier nuclei. For example if two deuterons (that is, two nuclei of that isotope of hydrogen of mass 2 which has the name deuterium[1]) collide together with a high enough energy (soon we shall see how great this energy has to be) one of the deuterons absorbs either the proton or the neutron from the other; in the first case a nucleus of the isotope of mass 3 of helium is formed, the neutron is freed and an energy of about 3·25 million electron volts is produced; there then occurs the following fusion reaction:

$$D^2 + D^2 \rightarrow He^3 + n^1 + 3 \cdot 25 \text{ Mev}$$

in the other case, in which one of the two colliding deuterons absorbs the neutron from the other, the result of the reaction is a nucleus of tritium (the isotope with mass 3 of hydrogen), a proton and about four million electron volts:

$$D^2 + D^2 \rightarrow T^3 + p^1 + 4 \text{ Mev}$$

In order that a fusion reaction should take place it is necessary that the two nuclei which take part (in the preceding example the two deuterons) come into intimate contact with each other. Now as we know, nuclei have a positive electric charge; thus they repel each other. It is therefore necessary to give them an energy which is great enough to overcome this electrostatic repulsion if we want them to come close enough to fuse together and form a new nucleus. And so the mass of gas in which one wants the fusion reaction to take place (in the preceding example the gas of deuterons) should be maintained at a very high temperature so that the nuclei which make it up have a velocity due to thermal agitation which is sufficient to overcome the coulomb repulsion.

Thus, in general, in order to trigger fusion reactions a certain amount of energy must be supplied. Naturally if the fusion

[1] We have already noted a deuteron is formed of a proton and a neutron.

Nuclear Energy

reaction is to be used for the practical purpose of producing energy it is necessary that the energy which has to be supplied in order that the reaction can take place should be *less* than the energy which it produces. This energy appears in the form of energy of motion of the reaction products and can be converted into energy which can be used industrially (for example into electrical energy).

It is not in all fusion reactions that the energy liberated is greater than the energy supplied; in practice this occurs only for the reactions between exceedingly light nuclei: the isotopes of hydrogen and of helium. Indeed as one goes to the heavier nuclei their positive electric charge increases and therefore there is a continual increase in the energy which we must give to them in order to overcome the mutual repulsive force.

Here then are some possible fusion reactions in addition to those we have already given:

$$D^2 + T^3 \rightarrow He^4 + n^1 + Q_1$$
$$D^2 + He^3 \rightarrow He^4 + p^1 + Q_2$$
$$D^2 + n^1 \rightarrow T^3 + Q_3$$

in which as usual D, T, He, stand for, respectively, the nuclei of deuterium (the deuteron), of tritium (the triton), and of helium (the alpha particle); n and p stand for the neutron and the proton and finally Q_1, Q_2, Q_3 are the amounts of energy which are liberated in the reaction: these are some ten times greater than the energy which must be supplied to the nucleus in order for the reaction to take place.

The last reaction is different from the others: in fact one of the two particles which react together is not in itself a nucleus but a neutron which, being electrically neutral, is not affected by any repulsive force from the other nucleus. This reaction (like others in which neutrons are involved) does not require therefore that energy should be supplied to the reacting particles. Unfortunately however, as we know, isolated neutrons do not exist in nature and if we wish to use them it is necessary to produce them by means of a nuclear reaction.

Let us exclude this last reaction and ask ourselves how much energy the reacting nuclei must possess if the fusion reaction is to take place. It is found that this energy must be about 20,000 electron volts. The atoms in a receptacle which is maintained at room temperature have approximately an energy of one-fortieth of an electron volt; the higher the temperature the higher is this energy of motion of the atoms. They would have an energy of 20,000 electron volts if the temperature were greater than 200 million degrees centigrade! Only if one were able to produce a gas, for example, of deuterons of this enormous temperature would it be possible for fusion reactions to take place. Then the very energy which is liberated in the first fusion processes would heat up the neighbouring gas and so would favour the production of other reactions.

We now understand why fusion reactions are also often called *thermo-nuclear reactions*.

Even before 1930 some physicists had found that protons and other light nuclei when accelerated in an accelerating machine to some thousands of electron volts and directed against a target made of a light element were able to overcome the electric repulsion of the target nuclei and to fuse with them giving rise to nuclei of heavier elements; i.e. they were able to produce fusion reaction. But to obtain one of these reactions it is necessary to expend an enormous amount of energy. There would have been hope of being able to use the energy liberated by fusion for practical purposes only if it had been possible to arrange that the reaction once kindled was self-maintaining.

This is what happens in a normal chemical combustion which once started at a given point in the fuel maintains itself, thanks to the heat which is developed in the reaction itself. In the fusion reactions, as we have seen, the situation is similar because these also produce a certain amount of energy which favours the production of further fusion in the particular element which can be considered as the fuel (for example, deuterium).

But between a chemical combustion and a fusion reaction there is an enormous difference: in fact in the first the ignition temperature has the value of some hundreds of degrees, but in order that a fusion reaction should take place it is necessary, as we have seen, that the temperature should reach the value of hundreds of millions of degrees.

PLASMA

At these enormous temperatures matter is not solid, not

7a The synchrotron at Frascati near Rome

7b Internal view of the ring of the 25 Gev proton synchrotron at Cern, Geneva

8 External view of the Ispra reactor

liquid and is not even gaseous in the ordinary sense of this word; a gas is a substance whose molecules have a kinetic energy such that each of them moves through the substance as if it were free, that is not subject to forces from the other molecules; it is deviated in its path only by collision against the other molecules and against the walls of the container.

At these very high temperatures matter exists in a special state which is called *plasma*: it is a special type of gas made up of nuclei and of free electrons. Indeed the energy is such that all of the satellite electrons are torn out of the atoms; they are reduced to bare nuclei. Plasma then is a gas which is made up of nuclei and of electrons in very rapid and chaotic movement. However it behaves, considered as a whole, as an electrically neutral system, just as does a normal gas. This is because the number of negative charges (electrons) which it contains is on average equal to the number of positive charges (nuclei).

Here is a gripping and dramatic account in which the physicist R. F. Post, in a paper entitled 'Fusion Power', describes an imaginary experiment which exhibits very well the strange properties of plasma:

'We take a litre of deuterium gas confined in a vessel made of a mythical material which is capable of withstanding the enormous temperatures and pressures that will arise in the course of the experiment. At room temperature and normal atmospheric pressure the deuterium gas-molecules are wandering about in the vessel with an average kinetic energy of about one twenty-fifth of an electron volt, or a velocity of about 3,000 miles per hour. Of course no fusion reactions are taking place. Now we heat the gas to 5,000 degrees C. At this temperature we no longer have molecules: the violence of their collisions has broken them apart into deuterium atoms. The pressure has risen to about 40 atmospheres (600 pounds per square inch), and the average velocity of the atoms is about 40,000 miles per hour. But we are still very far from the velocity needed to make the nuclei fire.

'Next, let us jump to 100,000 degrees. The remarkable properties of the mythical wall material are very much needed now, for any real material would long since have vaporized. Now the deuterium atoms of the gas have been broken down to the electrically charged nuclei (deuterons) and electrons: in a word, the gas has become what is known as a plasma. The gas pressure has risen to 1,500 atmospheres. The average velocity of the electrons is 10 million miles per hour, and even the much heavier deuterons are moving at the great speed of 170,000 miles per hour. Yet the deuterons still do not have sufficient energy effectively to overcome their mutual electrostatic repulsion. At this temperature there would be only about one fusion in the litre of plasma every 500 years! We still have a long way to go before we shall reach the ignition temperature of a mass of deuterons.

'At one million degrees, the rate of fusion reactions will increase more than a billion times, but the total energy output will still be too small to be detected—only a few millionths of a watt per cubic centimetre. At 100 million degrees, however, the reaction rate will become really respectable. The pressure then will have reached the staggering value of 15 million atmospheres. The electrons will be travelling at 90,000 miles per *second*, and the deuterons at 1,500 miles a second (around the world in sixteen seconds). Essentially all of the deuterons will react with one another rapidly (within a fraction of a second), and their reactions will release energy at a fantastic rate—about 100 million kilowatts. But we shall not yet have arrived at the kindling point: to sustain the reaction we shall still have to put in more energy than the fusions release. Only at about 350 million degrees will the "fire" (i.e. the thermonuclear reaction become self-sustaining.'[1]

This imaginary experiment shows exceedingly well what are the conditions which must be achieved in order that a fusion reaction should maintain itself and the terrible difficulties which face anyone who proposes to obtain one.

The enormous pressure reached by the deuteron gas suggests that it is not possible in any real experiment to start from the conditions from which the imaginary experiment was started and in which the gas had initially the normal atmospheric pressure. We must, on the other hand, start with a very much lower initial pressure: so low that in the laboratory it would be considered as almost a vacuum. This very low pressure leads to a strange situation: although the temperature, considered

---

[1] From 'Fusion Power' by Richard F. Post. Copyright © 1957 by Scientific American Inc. All rights reserved.

as the kinetic energy of the particles which make up the plasma, will reach exceedingly high values when the reaction becomes self-maintaining the gas will nevertheless contain little heat: a litre of the plasma of deuterium at the pressure of a ten thousandth of an atmosphere would contain, at the kinetic temperature of 350,000,000 degrees, only 18,000 calories which would be just enough to heat a small cup of coffee.

ENERGY LOSSES IN THE PLASMA

If then one starts with gas of low density there is no longer the risk of destroying the walls of the container as the temperature is increased. This does not however mean that we can safely raise to some hundreds of millions of degrees the plasma which is simply placed in a container; for the walls would produce a continuous cooling of the plasma and so inhibit the fusion reaction. This occurs for the following reason.

The particles of the gas, because of the extremely low density, have very great freedom of movement (at that terrible velocity to which I have referred above); and each of them could travel thousands of kilometres before meeting another. On the other hand at a very much higher frequency they meet another obstacle: the walls of the container; and in every one of these extremely frequent collisions each particle loses some of its energy. The plasma cools and the fusion reaction is checked.

What then? How can we keep the plasma in a container and at the same time prevent it from touching the walls? In the next section we shall see how this problem which appears insoluble is now being tackled.

But this is not the only means by which the particles which make up the plasma can lose energy; there is in addition the emission of light and the emission of braking radiation.

Although it may seem very unlikely very hot plasma emits very little light. The reason is obvious when one recalls the way in which light is emitted by matter. In normal conditions the atoms which make up matter are complete, that is, they have their troop of electrons which revolve around the nucleus; and light is emitted when this structure alters, when, that is, one of the more external electrons is shifted so that it revolves in a still more external orbit than that in which it previously revolved: subsequently it falls back into the original orbit and in doing so emits light. Thus an essential condition for the emission of light is that the atoms should possess their satellite electrons. But this condition does not obtain in a plasma; in a plasma the nuclei are stripped bare and the electrons are free. Light can be emitted only by those very few electrons which, having collided with a nucleus, have formed an atom (ionized molecule). But this exists only for a very short moment since due to the violent collisions the electron will very soon be stripped from its nucleus, its ephemeral companion, and will return to its erratic and solitary life.

But although the energy loss of the plasma as a result of light emission is small it suffers very great energy loss as a result of the braking radiation which is emitted in the form of X-rays. This radiation originates in the plasma from the violent collisions of the free electrons against the nuclei; as a result of these collisions the electrons are slowed down and therefore lose a part of their energy in the form of X-rays, that is, of electromagnetic radiation with very short wave-lengths. And as X-rays are extremely penetrating this energy will be dispersed through the walls of the container.

Therefore (if we exclude for the moment the cooling of the plasma which is produced by the walls of the apparatus which contain it) some of the energy produced in the fusion reactions which take place in the plasma will be lost because of the slowing down of the electrons. Clearly, if the fusion reactions are to propagate through the mass of the plasma, it is necessary that this lost energy should be less than that which is produced; it is necessary therefore to reduce to a minimum the collisions between electrons and nuclei.

Fortunately the number of collisions is reduced as one reduces the density of the gas; they become relatively rare at that density of plasma at which we have decided that it is convenient to make the experiment: that is, at about one ten-millionth of an atmosphere.

On the other hand in order to increase the energy produced in fusion it is necessary to increase the temperature. One finds in this way for each particular plasma a *kindling temperature*, that is a temperature such that for the particular plasma the energy produced is adequate, despite the losses, for the propagation of the reaction throughout the entire mass. In the case of the reaction deuteron-triton this ignition temperature is about 50 million degrees; on the other hand, for the reaction deuteron-deuteron a temperature greater than 350 million degrees is needed.

THE MAGNETIC BOTTLE

But let us return to that problem which might appear insoluble: how to hold the plasma in a container and yet prevent it from touching the walls? About ten years ago an ingenious solution to this problem was suggested.

As the walls of the container cannot be made of a material substance (because this would already melt at a temperature of some thousands of degrees) it was suggested that the plasma should be confined in a magnetic field which would 'imprison' it and prevent it from touching the walls. This is possible because the plasma is made up of electrically charged particles.

A strong magnetic field obliges all the charged particles which move in it to travel on a circular path. Therefore if the plasma in its container is surrounded by a strong magnetic field the particles which make it up are trapped by this invisible magnetic wall and cannot reach the walls of the container where they would cool themselves. It is found that a magnetic field of 500,000 gauss (which can be obtained in the laboratory) is able to trap a plasma in which the pressure is 10,000 atmospheres.

Unfortunately, the behaviour of this *magnetic bottle* is not as simple as might appear from what I have said. The study of the interactions between a magnetic field and a very high temperature plasma has opened a new unexplored field of physics which still presents many unknown features.

THE PINCH EFFECT

But how is it possible in practice to obtain a magnetic bottle? Different methods have been suggested and various arrangements have been constructed and tested by physicists of different countries in a research effort which demands the most advanced resources of modern technique.

Amongst the various methods which have been suggested I

want to emphasize one which is extremely elegant and which is based on what is called the 'pinch' effect. This is what it is.

When an extremely intense current passes through a conducting gas which is contained in a tube it gives rise to a magnetic field whose lines of force encircle the gas and compress it away from the walls of the tube (Fig. 37).

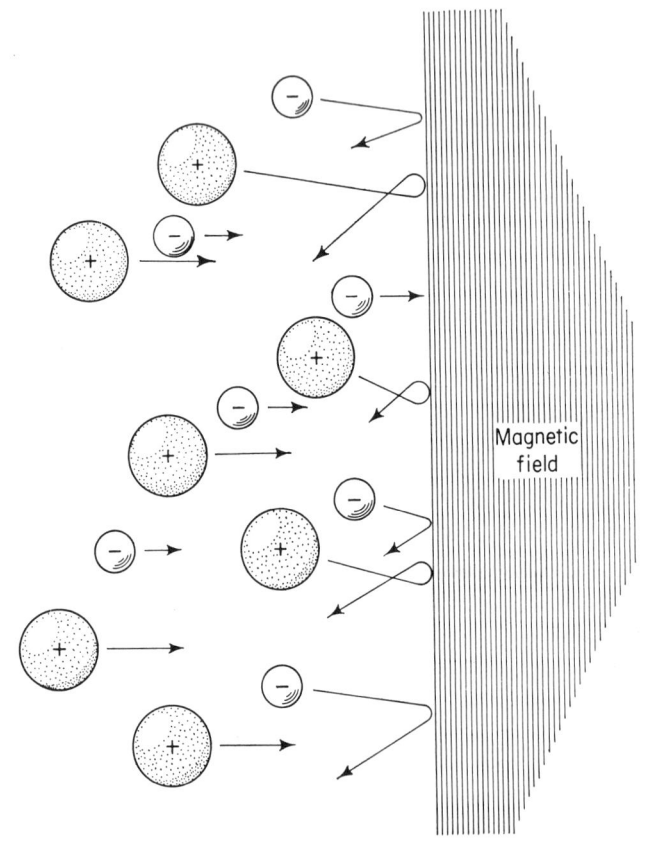

Fig. 37. The pinch effect.

Here then is a means already available for making a magnetic bottle for a plasma which is a very good conductor of electricity. As usual, however, many difficulties arise in the practical application: there are many sources of instability so that the time for which the effect persists is exceedingly short; about one-millionth of a second. Naturally various systems are being studied in an effort to eliminate or at least to reduce these sources of instability.

In the designs which are used for these studies we find that the tubes which contain the plasma are not straight but are bent in various forms: oval, toroidal, S-shaped, etc. (Fig. 38). Indeed, amongst other things, in a straight tube the electrodes at the two ends would produce a cooling of the plasma in contact with them.

Some of the arrangements constructed for the study of fusion make use of the pinch effect. Others start with a cold gas contained in a magnetic field and heat it by means of electrical discharges. Others again speed up the ions with an accelerating mechanism and then imprison them, for example, in a rapidly increasing magnetic field.

Research on fusion is carried out by scientists throughout the world. The day will probably come in which man will know how to exploit the enormous energy liberated in fusion reaction; but perhaps that day is not as close as some believe and as everybody hopes.

THE ENERGY OF THE STARS

The ultimate end at which these investigators hope to arrive is the realization of an apparatus (a *fusion reactor*) in which the fusion reaction, once kindled in the fuel made with light nuclei (deuterons or tritons or helium), succeeds in becoming self-maintaining and supplying in a controlled manner energy which once extracted can be used for industrial purposes; they propose that is to construct what has been called 'a piece of sun'.

Indeed the enormous energy which the stars (and in particular the sun) continuously radiate in the form of light and of heat originates in the nuclear reactions of fusion which take place inside them.

For a long time men have asked themselves how a star is able to emit for millions and millions of years such an enormous quantity of energy. For example, the sun is not a particularly large star and is not very hot but it irradiates energy of about 400 billion, billion watts. However, until about thirty years ago

(a)

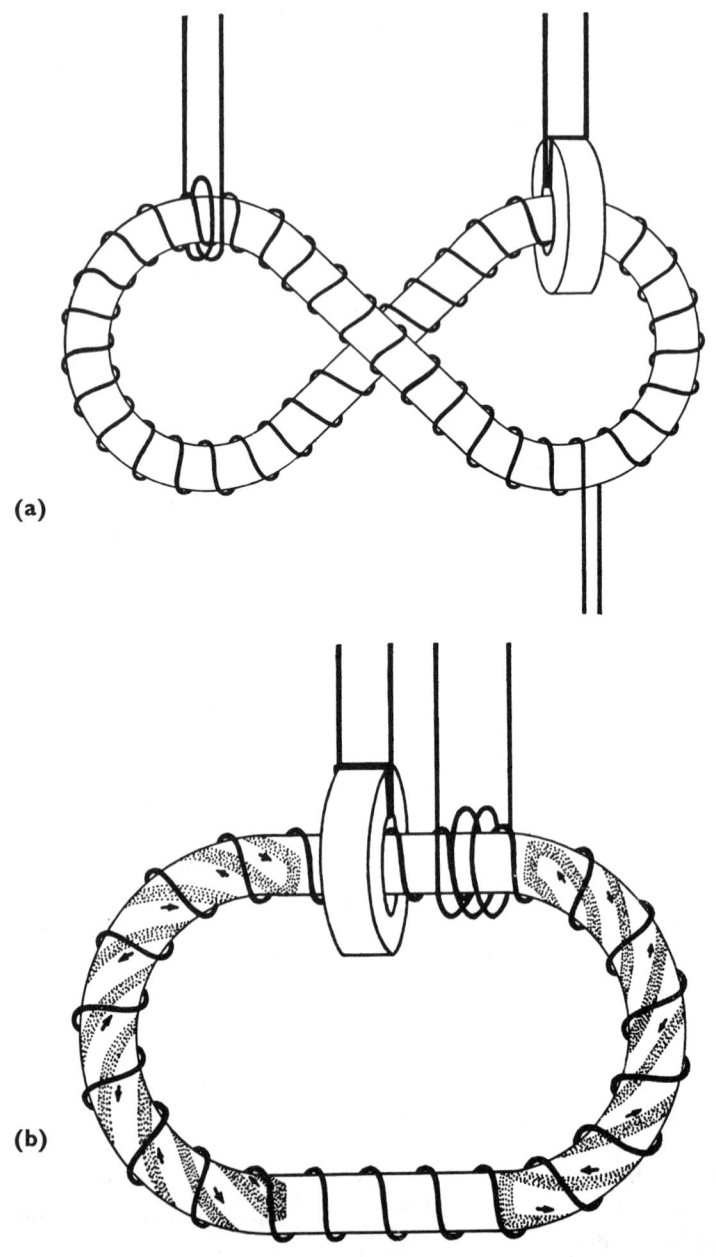

(b)

Figs. 38 (a) and (b). Types of magnetic bottle used in the study of fusion reactions.

Nuclear Energy

all the attempts which had been made to explain this grandiose phenomenon had failed. It was only in 1930 that the research of the English astro-physicist, Eddington, and others on the physical conditions existing inside stars supplied the key which eight years later opened the solution of the problem.

Stars are enormous masses of extremely hot gas, consisting essentially of hydrogen, whose internal temperature reaches a value in the range between 15 and 30 million degrees. At this extremely high temperature all the atoms are completely ionized; and the nuclei and the electrons which make up the gas have sufficient energy to overcome the electrostatic forces of repulsion and to give rise to nuclear reactions of fusion. We have then a plasma and we are in conditions such that fusion reactions not only take place in this plasma but are also self-maintaining. For the huge mass of very hot gas which surrounds the inside of the star prevents it from losing too much heat by radiation. If we could make a fusion reactor as large as the moon, we should not have to worry overmuch about those losses of energy to which I have referred in the previous section and which constitute perhaps the most serious obstacle to the realization of one of these exceedingly small fusion reactors which we can construct on this earth of ours.

When astro-physicists were certain that stellar energy orginated in fusion reaction there arose the problem of finding, amongst all the possible reactions, those which take place at a temperature existing in the inside of stars; the reactions had moreover to take place with the production of the appropriate energy and to be made by the nuclei of those elements which we know to be present in sufficient quantities in the stars.

The following conclusion was reached: there were two types of fusion reaction responsible for the production of energy in the greater part of the stars; each of these in the long run consumes only nuclei of hydrogen (that is they consume only that element which is by a very long way the most abundant of the elements found in the stars); and in each the products of the reaction are nuclei of helium.

The first type of reaction starts from the direct combination of two protons and is called the *proton-proton process*; the second on the other hand requires the presence of nuclei of carbon; these however act only as catalysts because they are regenerated at the end of the process: for this reason, this type of reaction is called the *carbon cycle*.

The proton-proton process is the following:

$$p^1 + p^1 \to {}_1D^2 + \epsilon^+$$
$$_1D^2 + p^1 \to {}_2He^3 + \gamma$$
$$_2He^3 + {}_2He^3 \to {}_2He^4 + 2p^1$$

As we see, in this process two protons combine to yield a deuteron and a positron; according to the calculations of Bethe and others, in the temperature and pressure conditions which exist in the sun, two protons interact in this manner on average once in seven thousand million years. Hardly ten minutes after its formation the deuteron captures a proton, forms a nucleus of helium of mass 3 and liberates a gamma quantum; then, in about 300,000 years two nuclei of helium 3 combine to form a nucleus of helium with mass 4 (that is an alpha particle) and two protons.

The carbon cycle is more complex; in it the following reactions are involved:

$$_6C^{12} + {}_1p^1 \to {}_7N^{13} + \gamma$$
$$_7N^{13} \to {}_6C^{13} + \epsilon^+$$
$$_6C^{13} + {}_1p^1 \to {}_7N^{14} + \gamma$$
$$_7N^{14} + {}_1p^1 \to {}_8O^{15} + \gamma$$
$$_8O^{15} \to {}_7N^{15} + \epsilon^+$$
$$_7N^{15} + {}_1p^1 \to {}_6C^{12} + {}_2He^4$$

As we see the nuclei of carbon are not consumed in this process: for every carbon nuclei which starts the cycle another is produced at the end of the cycle.

Different temperatures are required for these two processes to take place. The theory shows, for the exceedingly large class of stars which belong to the same type as the sun, that the carbon cycle forms the main source of energy when they have an internal temperature higher than about 20 million degrees; whilst for stars with internal temperature of about 16 million degrees the two processes contribute about equal amounts of energy; finally for the stars with internal temperature of about 13 million degrees the major contribution is supplied by the proton-proton process.

By this means the sun changes every second 564 million

tons of its hydrogen into 560 million tons of helium; i.e. in every second 4 million tons of material are converted into energy.

FISSION

Whilst in the fusion process one obtains energies by starting from extremely light nuclei and making nuclei which are less light in the *fission* process one obtains energy by breaking up very heavy nuclei. Until now we have been concerned with fusion and we have made a rapid review of the state of research today. We will now turn our attention to fission, that is to that particular type of nuclear reaction which in practice appears only in nuclei of very high atomic number: thorium ($Z = 90$), protactinium ($Z = 91$), uranium ($Z = 92$), and successive elements which although not existent in nature can be produced artificially.

A nucleus of one of these heavy elements, for example a nucleus of uranium, when struck by a neutron (or by another projectile) absorbs it and immediately breaks up into two other nuclei of intermediate atomic weight together with two or three neutrons. This is the process of fission. However, a nucleus when struck by a neutron can give rise to different fission reactions according to the way in which its constituent nucleons, together with the projectile neutron, rearrange themselves amongst the two fragments and the neutrons which are liberated. For example a nucleus of uranium, of atomic weight 235, when struck by a neutron can give rise to a nucleus of technetium ($Z = 52$), a nucleus of zirconium ($Z = 40$) and to three neutrons; or on the other hand to one of scandium ($Z = 50$) to one of molybdenum ($Z = 42$) and to four neutrons, etc. If a mass of uranium is bombarded by neutrons there occur in it in various proportions all of the possible fission reactions giving rise to about eighty isotopes of nuclei with intermediate atomic weight.

These different nuclei which are produced in fission reactions from nuclei of uranium have, however, one characteristic in common; they possess too many neutrons to be stable; in the neutron-proton diagram (Fig. 27) they are to be found above the region in which are grouped the stable nuclei. The reason for this is very simple.

The neutron-proton diagram shows that the points which represent all the stable nuclei are concentrated along a line which at the origin coincides with the bisector of the axes but which as it rises becomes displaced from the bisector; this means that whilst the light nuclei normally contain a number of neutrons almost equal to the number of protons (Z equal to roughly A–Z, that is A approximately to $2Z$) as one moves to heavier nuclei the number of neutrons in the nucleus becomes greater than the number of protons (A becomes equal to approximately 2·6 Z).

For this reason when a nucleus of uranium absorbs a neutron and breaks into two nuclei of intermediate atomic weight one or other of these will have a number of neutrons greater than that which it would have if it were stable: the point which represents it in the neutron-proton diagram lies above the line of stable nuclei. It will therefore be an unstable nucleus which by successive emission of electrons (which produce an increase in the atomic numbers Z and a reduction in the number N of neutrons) reduces itself to a stable nucleus; that is the unstable nucleus will be the head of an artificial beta radioactive family.

This is the reason why uranium when irradiated with neutrons produces that complex radioactivity which had been observed in 1934 by Fermi and his collaborators during the first experiments on the bombardment of uranium by neutrons.

CHAIN REACTIONS

In the fission process there are emitted a certain number of neutrons. This fact is of very great importance; it provides the means for propagating the reaction to all the nuclei in a mass of *fissile substance* (i.e. a substance capable of giving rise to fission).

A nucleus in a mass of uranium, when struck by a neutron, breaks up and emits two or three neutrons; each of these neutrons may trigger a fission of another nucleus with the production of new neutrons which in their turn will trigger new fissions and new neutrons . . . and so on. One will have what is called a *chain reaction*, thanks to which one single initial neutron triggers the fission of an enormous number of nuclei.

When I spoke of fusion reactions I noted that in these as in chemical combustion the reaction spreads itself throughout the mass thanks to the heat developed in the reaction itself. It is for this reason that fusion reactions are also called 'thermo-nuclear reactions'; for in them it is heat that is responsible for the propagation of the reaction.

In fission, on the other hand, it is not heat that is responsible for the propagation of the reaction through the entire mass of fissile substance; this is done by the neutrons.

The possibility of kindling a chain reaction in a fissile substance (i.e. in the final analysis the fact that in fission, which is produced by neutrons, a certain number of neutrons are produced) has made possible the production of apparatus which permits the practical exploitation of nuclear energy.

When a nucleus of uranium 235 is struck by a neutron and breaks up it liberates an energy of about 190 million electron volts. This means that the fission of all the nuclei which make up a kilogram of uranium 235 would set free an energy of 19,000,000,000 calories. A kilogram of coal would produce when burnt 7,500 calories. Thus one *kilogram* of uranium 235 would develop when it breaks up the same energy as one would obtain by burning about 2,500 *tons* of coal.

URANIUM

Uranium is a silvery white metal which is not found free in nature but is however found in compounds which are fairly common. It was discovered in 1789 and was given its name in honour of the planet Uranus which had been discovered shortly before. It is contained in certain minerals such as pitch-blende (the principal deposits of this are in Canada, in the Congo and in Bohemia) in carnotite (which is found in Colorado, USA) and in autunite.

Natural uranium is a mixture of three isotopes of masses 234, 235 and 238; all of these are radioactive with very long half-lives; they break up yielding an alpha particle and a gamma ray.

In natural uranium which is extracted from these minerals the three isotopes exist in very different proportions; much the most abundant is uranium 238. The actual proportions are given below:

	%	half life
U^{234}	0·006	240,000 years
U^{235}	0·715	710 million years
U^{238}	99·279	4,590 million years

So natural uranium consists essentially of U^{238}.

Let us ignore U^{234} which forms only a very small fraction of natural uranium and let us see how the two different nuclei U^{235} and U^{238} behave when they are bombarded with neutrons. It has been found that the fission of uranium nuclei can be produced both by fast neutrons and by slow neutrons. The details of the interactions are as follows:

(1) Slow neutrons produce the fission of U^{235} but not of U^{238};

(2) Fast neutrons are less effective than slow neutrons in producing the fission of U^{235};

(3) Fast neutrons, provided the energy is sufficiently high, also produce the fission of U^{238};

(4) Slow neutrons are captured by U^{238} which undergoes a (n, γ) process which leads to the following sequence of transformation:

$$_{92}U^{238} + {_0}n^1 \rightarrow {_{92}}U^{239} + \gamma$$
$$_{92}U^{239} \xrightarrow{\text{23 minutes}} {_{93}}Np^{239} + \epsilon^-$$
$$_{93}Np^{239} \xrightarrow{\text{2·3 days}} {_{94}}Pu^{239} + \epsilon^- + \gamma$$
$$_{94}Pu^{239} \xrightarrow{\text{24,000 years}} {_{92}}U^{235} + {_2}He^4$$

Thus a nucleus of U^{238}, when it is struck by a slow neutron, absorbs it and is changed into another isotope of uranium (which does not exist in nature): U^{239}. This with a mean life of 23 minutes is transformed emitting an electron into an element which does not exist in nature, neptunium, (Np) of atomic number 93 and atomic weight 239; finally neptunium with a mean life of 2·3 days is transformed into another element which again does not exist in nature, plutonium (Pu) of atomic number 94 and atomic weight 239, again emitting an electron and a gamma ray.

The nucleus of plutonium is virtually stable; it is transformed into U^{235} with a mean life of 24,000 years.

It is important to note that plutonium when bombarded with neutrons behaves like U^{235}, i.e. it undergoes fission with the emission of neutrons even if the projectile neutrons are slow.

Nuclear Energy

The first transformation—i.e. the process U^{238} (n, γ) U^{239}—and the consequent formation of plutonium occurs with very high probability for slow neutrons of a certain well-defined energy; we have here the case of a nucleus which has absorption lines of the type which were explained on pages 205 and 206.

Let us now consider a mass of natural uranium and suppose that a nucleus of U^{235} inside this mass is broken apart. Four different things may occur to the neutrons which are emitted in this process:

(1) they may be captured by some impurity which is present in the mass of uranium;

(2) they may escape from the mass;

(3) they may be captured by nuclei of U^{238} without producing fission;

(4) they may finally be captured by U^{235} and produce fission.

For simplicity I have neglected the capture by U^{238} with the production of fission because the probability of this process is extremely small for bombarding neutrons of energies equal to or less than that of the neutrons which are emitted in the break-up of U^{235}.

Clearly a necessary condition for the chain reaction to be self-maintaining is that the losses due to the first three causes should be reduced to a minimum. Neutrons on the other hand which suffer the fate listed under No. 4 above are just those which can be used both to yield energy and to produce other neutrons destined to provide for the further maintenance of the chain reaction.

In order to reduce the number of neutrons captured by impurities present in the mass very careful purification is required.

The percentage of neutrons which escape from the mass of uranium and which therefore make no contribution to fission can be reduced by appropriate modification of the dimensions of the mass. Clearly the greater the volume of this mass the smaller is the probability that the escape of neutrons will be more important than capture leading to fission. The *critical size* of the mass of uranium is defined as the size for which the number of neutrons produced in fission is equal to the number of neutrons which escape from the mass or which are captured without producing fission. The value of this critical size depends on the composition of the mass containing uranium. A fission reaction will be spontaneously self-maintaining only if the size of the mass is greater than the critical size.

It is very difficult to reduce sufficiently the number of neutrons which are lost according to point (3) above, i.e. the number of neutrons which are captured by U^{238} according to the process discussed on page 271. This is obtained by suitably adjusting the composition of the mixture, that is by suitably reducing the percentage of uranium 238.

A fission chain reaction can be used for two different purposes:

(1) to obtain a controlled reaction in order to produce continuously energy which can be used industrially (*nuclear reactors*);

(2) to obtain the use in war of an explosive reaction (*atomic bombs* or more appropriately *nuclear bombs*).

NUCLEAR REACTORS

The machines which use fission chain reactions for industrial purposes were called initially *atomic piles*; today they are called *nuclear reactors*.

On December 2, 1942, the first nuclear reactor started working at Chicago; it was made under the general direction of Fermi assisted mainly by a group of physicists and technicians directed by Zinn and Anderson. On that day for the first time in the history of the world man succeeded in setting in operation a self-maintaining nuclear reactor.

At first the reactor was made to operate at a power of $\frac{1}{2}$ watt; a few days later this was raised to 20 watts. In May 1944 the power of 18,000 kilowatts was reached in a pile made at Clinton. Since then the power achieved has been further increased as I will describe below.

The principle of operation of a nuclear reactor is fairly simple. A slow neutron causes the break up of a nucleus of U^{235} which, as it breaks up, also emits two or three fast neutrons. In order that these neutrons can produce the fission of other nuclei of U^{235} they must be slowed down; to do this a *moderator* element is required; the moderator, which is suitably mixed (and we shall see immediately how) with the uranium, slows down the fast neutrons emitted in the fission before they collide with other nuclei of U^{235}. This slowing down occurs thanks to the elastic collisions which the fast neutrons undergo against the atoms of the moderator.

To be efficient a moderator must have two characteristics:

(1) it must slow down the fast neutrons in a small number of collisions and therefore its atoms must have a mass which is not very large compared to the mass of the neutron;

(2) the nuclei of its atoms must have a very small probability of capturing the neutrons which collide with them because otherwise many neutrons would be lost.

The moderators which have been found to be most efficient are: heavy water, beryllium, beryllium oxide, and graphite. Normal hydrogen, which would have the advantage of slowing down the neutrons better than any other nucleus, is not suitable because it would capture rather too many neutrons.

So the moderator slows down the fast neutrons (of about a million electron volts) until their energy reaches the value of a fortieth of an electron volt (slow neutrons). Unfortunately, however, the energy of the neutrons which are captured by nuclei of U^{238} without fission has a value which is intermediate between the energy of the neutrons emitted in the fission of U^{235} and the energy of the slow neutron; it is some tenths of an electron volt.

If, therefore, one were to construct a reactor by mixing together uniformly the moderator with the natural uranium during the slowing down process the neutrons would pass through this resonance energy and would be captured by the nuclei of U^{238}; thus they would be useless for maintaining the chain reaction and the reactor would 'converge'.

Now let us study a particular arrangement made to overcome this snag; we must mix the uranium with the moderator in such a way that the greater part of the fast neutrons which are produced in the fission of U^{235} are slowed down to a value lower than that which corresponds to the absorption lines of U^{238} before they encounter the nuclei of this isotope. The solution to this problem was suggested by Fermi and Szilard.

Instead of mixing together uniformly throughout the reactor

the natural uranium and the moderator they thought of disposing the uranium inside the mass of the moderator in the form of many parallel and equally separated cylinders, or many spheres arranged according to a grid pattern. If the distance between the elements is suitably chosen a fast neutron which is emitted in the fission of a nucleus of U^{235} contained in one of the bars of the grid will pass into the moderator and, before going on to strike another bar of the uranium, will have reached the energy of one-fortieth of an electron volt. It will now no longer have the energy appropriate to being captured without fission by U^{238}; it will have the energy appropriate for producing the fission of U^{235}.

This provided an elegant solution to the problems of reducing to a minimum the loss of fast neutrons due to the capture without fission by U^{238}.

One cannot with safety operate a nuclear reactor (Fig. 39) without setting up a control system which will allow one to slow down at will the operation or even to stop it completely.

The control of the reactor may be obtained, for example, by introducing into suitably prepared apertures in the mass of the moderator some bars of steel and cadmium. The cadmium captures the slow neutrons which are present in the pile so that by introducing or drawing out these bars one can vary at will the number of neutrons 'lost' from the process of maintaining the chain reaction.

In the starting conditions, with the reactor switched off, the bars of cadmium will all be fully inserted into the pile. In order to start the reactor working they will be slowly withdrawn; at a certain point conditions will be reached in which the chain reaction diverges and the reactor will start to produce a power which tends to increase. If this exceeds a certain safety limit the steel and cadmium bars are automatically reintroduced into the apparatus by the control system. We can now understand how the power produced by a reactor can be controlled.

Finally during the operation of a reactor there are two further important problems. First of all a reactor produces a very large amount of heat: it is therefore necessary to cool the mass in order to maintain the temperature below the limiting value at which the reactor can operate without the melting or damage

Fig. 39. Sketch of a nuclear reactor.

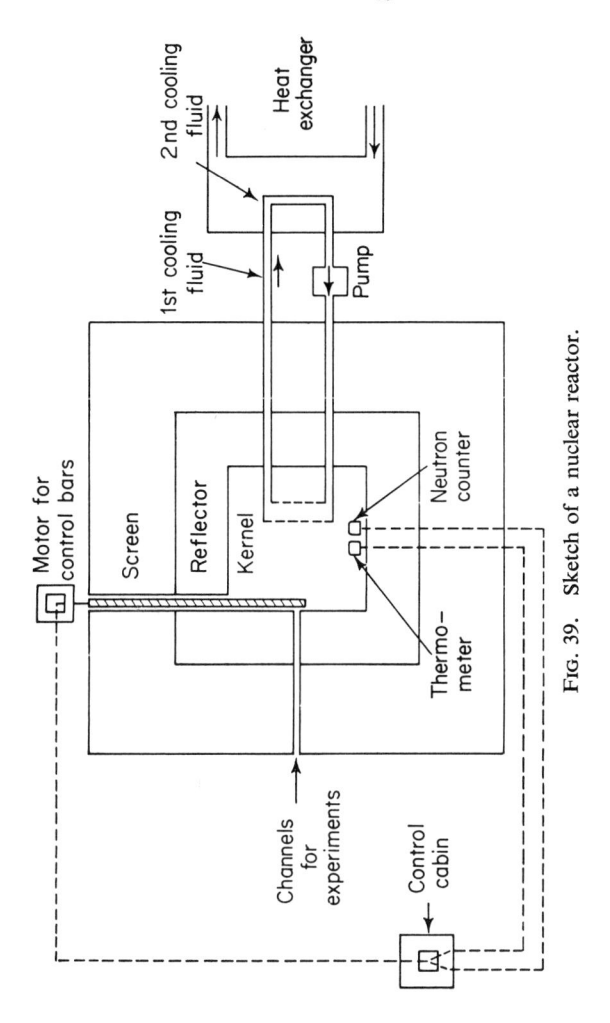

of one of its parts. This cooling is made by blowing a suitable gas (helium, oxygen, argon, ...) through ducts which are prepared during the erection of the reactor.

In the second place the secondary reactions which take place in the reactor operating at full power are so great that it is impossible for the control personnel to come near it. The whole system of the reactor therefore must be closed inside very thick walls of cement or of other absorbing material; all the control manoeuvres and maintenance of the plant must be carried out through this protecting screen.

These and other innumerable technical problems were resolved in the feverish work carried out in America by physicists and by technicians in the three years between 1941 and 1944 with an effort and organization which had no parallel in the story of mankind.

TYPES OF REACTOR

In the last few years enormous progress has been made in the design and construction of nuclear reactors.

They can be divided into three large classes according to the purposes for which they are required.

(1) *Research reactors* are used to carry out experiments on neutrons, on radioactive isotopes and for the study of the behaviour of various materials when exposed to intense radiation by neutrons and by gamma rays.

(2) *Power reactors* are made to produce energy. They yield in the form of heat an energy which is of the order of some thousands of kilowatts; this heat is used in a boiler to produce steam which then produces mechanical work in a turbine; this work is either used directly (for example to propel a ship) or is transformed by means of alternators into electrical energy.

(3) *Production reactors* serve for the large-scale production of plutonium, an element which as we shall see is used in the construction of atomic bombs.

ATOMIC FISSION BOMBS

Nuclear apparatus of an explosive character differs considerably from reactors.

One need only remember that in general an explosion consists in the production in a very short time and in a very restricted region of space a pressure and a temperature which are so high as to give rise to what is called a *shock wave*; this spreads itself in the surrounding space and destroys or damages by mechanical or thermal effect all that it encounters. Now, if one were to use as a nuclear explosive apparatus a system analogous to the reactor, before the temperature could reach some thousands of degrees the reactor itself would break up and cease to operate.

The time required for the slowing down of the neutrons (which is of the order of ten microseconds) is too long.

In general one can say that the higher the temperature and the pressure which one wishes to achieve in an explosion the more rapid must be the process which leads to this temperature and pressure, in order to avoid the disruption of the apparatus before the required conditions are reached.

To this we may add that a reactor would be too large and clumsy to be used for military purposes.

One has therefore to obtain fission not with slow neutrons as in a reactor but with fast neutrons. However the fast neutrons which are emitted from the fission of a nucleus of U^{235} have an energy which reaches about a million electron volts; and neutrons of this energy can produce fission in the other nuclei of U^{235} but they are not energetic enough to produce fission of U^{238}. If one wants therefore to maintain a chain reaction with fast neutrons the bomb should be made only with U^{235}. Alternatively, since as far as fission is concerned the properties of plutonium are similar to those of U^{235}, one could construct a bomb using only plutonium.

There are various methods of separating U^{235} from U^{238}; they are based on the use of special mass spectrographs or on centrifugal or gaseous diffusion processes. But when one thinks that before 1940 no large scale separation of isotopes except hydrogen had ever been made one realizes the enormous amount of work that was necessary in order to produce the large amount of pure U^{235} required for the construction of an atomic bomb.

Plutonium as we have seen is produced in a reactor as the final product of transformation of U^{238}. It is separated with special and delicate chemical methods; these are complicated by the fact that the dangers due to the radiation (which demands the screening of the reactor) are also largely present in the plant for the separation of plutonium from the different fission products. For this reason all the stages of the separation plant must be screened and must be manipulated from a distance by a system of tele-control.

If one were to have an amount of U^{235} or of pure plutonium which was greater than the critical size there would be no way of preventing an immediate explosion; for there are always present a certain number of neutrons capable of kindling the chain reaction; these neutrons come from cosmic rays or from some spontaneous nuclear break-up or from reactions produced by alpha particles.

To prevent such an explosion it is therefore necessary that the bomb should be made up of two separate pieces, each of them smaller than the critical size. In order to produce an explosion it is necessary to bring together these two parts sufficiently rapidly to avoid the start of a chain reaction during the approach, that is before the bomb reaches its maximum efficiency.

Various methods can be used in order to produce this exceedingly rapid approach of the two parts of the bomb. One of these consists in using one of the two parts as a target against which the other part is fired. Another method consists in surrounding a sphere of fissile material of less than the critical size with a layer of explosive; the explosion of the layer of explosive compresses the fissile material so that its density is increased until it reaches the value at which the chain reaction can be kindled.

The first experiments on nuclear explosion was made in New Mexico on July 16, 1945. The first atomic bomb was dropped by the United States on Hiroshima (Japan) on August 6, 1945; it was equivalent in energy to 22,000 tons of high explosive. The second bomb was dropped by the United States on Nagasaki (Japan) on August 8, 1945.

After the end of the Second World War very many other atomic bombs were exploded by various countries for experimental purposes.

Hydrogen or H-Bombs. I do not want to discuss atomic bombs in too much detail; they are not looked on with favour by physicists (or indeed by any men in whom the thirst for power has not smothered a sense of responsibility) and very little is known about them because of military secrecy. But I will give a brief account of the so-called *H-Bomb* or *Hydrogen Bomb*.

In the bomb described above the process of fission of heavy nuclei (uranium 235, or plutonium) is used to obtain the instantaneous liberation of enormous amounts of energy; in the H-Bomb on the other hand the process of fusion of light nuclei is used. Thus both types are atomic bombs (or more precisely nuclear bombs); the name *A-Bomb* has been introduced to describe the explosive device which exploits the fission phenomenon and that based on the fusion of light nuclei is called the *H-Bomb*. It is known that an H-bomb uses a fusion reaction of tritium; for example

$$D^2 + T^3 \to He^4 + n^1$$

or

$$T^3 + T^3 \to He^4 + 2n^1$$

The exceedingly high temperature which is needed for the reaction to take place is supplied by the explosion of a fission bomb. An H-bomb produces an energy which is many hundreds of times greater than that produced by a fission bomb; every pound of explosive would produce about a hundred million kilowatt hours! But the power of an H-bomb can be still further increased by surrounding it with a layer of natural uranium (which as we know consists essentially of uranium 238); the neutrons emitted in the fusion reaction have a velocity such that they can produce fission of the nuclei of uranium 238 and this in its turn accelerates the fusion reaction.

EFFECTS OF ATOMIC EXPLOSION

The damage produced by normal chemical explosives is almost entirely due to the propagation in the air of a pressure wave produced by the sudden expansion of the explosive and by the splinters which are projected in all directions. The damage caused by the high temperature is relatively small because it is concentrated in a small space in which the mechanical damage is very much greater.

In the case of atomic fission bombs on the other hand, in addition to the effect of the pressure wave, there is a very outstanding effect due to the irradiated heat. In the vicinity of the explosion the temperature reaches the value of some millions of degrees in an extremely short time; as a result, there is emitted extremely intense electromagnetic radiation of all frequencies (light, X-rays, and gamma rays); this produces fires up to a distance of some kilometres and an ascending current of air.

In addition there are radioactive effects, by far the most lethal of which are due to the gamma rays emitted during the fission.

The experts affirm that the hydrogen bomb can liberate an energy one thousand times greater than that liberated by the A-bomb at Hiroshima; everything within a range of ten miles would be destroyed; everybody would be killed within a range of twenty miles.

Nobody who believes in the liberty and in the dignity of man, who has faith in the sincerity of relations between men and between the nations, who recognizes the value of the personality of the individual can conceive of the possibility of an atomic war in which the whole of our civilization would be destroyed. The use of nuclear arms is not a military or a political problem; it can only be considered as a moral problem.

THE PEACEFUL USES OF NUCLEAR ENERGY

Let us now make a very rapid review of the innumerable peaceful applications of nuclear energy.

First and certainly the most important is the production of energy. As I have said the heat of a reactor can be extracted by circulating within it a suitable gaseous or liquid substance which can in its turn operate a turbine. In order to transmit this energy elsewhere it is naturally convenient to transform it into electrical energy.

There are now in operation throughout the world more than 200 reactors; few of these are however power reactors, that is reactors which are used to produce energy. Others of various types are being designed and constructed. Only experience can tell us which type will be economically the most convenient.

The first plant for very large-scale production of electrical energy of nuclear origin was the power station of Calder Hall in Cumberland which started working in October 1956. It is composed of four identical units each of which consists of a reactor which supplies four turbo-alternators. Each reactor produces 42,000 kilowatts of electrical power. The nucleus of the reactor is contained in a steel cylinder 40 feet in diameter, 60 feet high and 2 inches thick; the nucleus is made of 1,150 tons of graphite and is perforated by 1,698 vertical canals which contain the fuel elements (uranium) and 40 canals which contain the regulating rods (boron steel) which have a stroke of about 20 feet. One of the characteristics of the reactors of this power station is that of being cooled by circulation of carbon dioxide at a pressure of seven atmospheres.

The first power station built in America started working in 1958 at Shippingsport in Pennsylvania and produces thermal power of about 60,000 kilowatts.

The first application of nuclear energy to propulsion was in the American submarine *Nautilus*. More than a hundred yards long and weighing 2,800 tons with a crew of ninety-five, it is the realization of the dream of Jules Verne. It can remain submerged for two months. It can travel (and already has travelled) under the polar ice; it can make several circuits round the world without surfacing.

The application of nuclear propulsion to merchant ships has already been realized. The first American merchant ship is called *Savannah* in memory of the ship (steam-propelled) of the same name which was the first ship flying the American flag to cross the Atlantic; it is about 200 yards long, 30 yards wide and displaces at full load 21,800 tons. It was launched from the Camden yards on July 21, 1959.

A nuclear reactor is not only a source of energy but also the most important source available to physicists of neutrons for use in experiments which could not be carried out by other means. Furthermore a reactor produces a very large quantity of isotopes, many of which are radioactive and do not exist in nature.

These artificial substances have been applied in every field of science.

The applications of radio isotopes in biology and in medicine are now extremely extensive; they are used both for research and for clinical investigation and therapy; in particular in diagnosis they are used for the localization and the diagnosis of tumours, for diagnosis in orthopaedic surgery and for the diagnosis and the evaluation of failings and of malfunctionings of the heart.

Artificial radioactive substances are also used in the treatment of cancer. I have already explained how this illness can often be cured by the application of preparations of radium or of emanation whose beta and gamma radiation have a destructive effect only on unhealthy cells. Substitution of artificial radioactive substances for natural radioactive substances is not however considered as economically convenient because of the very high cost of the plants for the production of radioactive substances. But in many cases these possess certain advantages over radium; indeed as the mean life of radium is very long it cannot be left inside the human body because after destroying the tissues of the tumour it would gradually destroy also the healthy surrounding tissues; on the other hand, if an artificial radioactive substance with a fairly short mean life is applied it can be without danger left inside the organ since it will have decayed before being able to damage in any way the healthy tissues.

Artificial radioactive elements are often used as *indicators* or *tracer elements*. One might for example want to study the metabolism of phosphorus in animals or in plants. All that is necessary is to supply to the organism in the required form a preparation in which even an exceedingly small fraction of the stable phosphorus is replaced by radioactive phosphorus; the latter, as far as all chemical and biological effects are concerned, is identical with normal phosphorus. This radioactive phosphorus—just because it is radioactive—can be followed allowing one to determine in a quantitative manner what percentage of the supplied phosphorus becomes fixed after a given time in one organ rather than in another. A vast amount of research of this type has been made not only with phosphorus but also with other chemical elements which are particularly important in the vital processes of plants and animals.

In the field of chemistry not only have reactors supplied elements which do not exist in nature but also the use of radioactive isotopes has led to the solution of many problems and to the posing of new problems: for example, there have been studied processes of polymerization, the method of formation of electrolytic deposits, the exchanges between a metal and the solution of one of its salts, the phenomenon of corrosion of materials, etc.

But this is not the place to give a list of the applications of nuclear energy and in particular of radio isotopes: in biology and in geology, in engineering and in agriculture, in all fields they constitute a new and extremely powerful means for study and for research.

CHAPTER ELEVEN

SUB-NUCLEAR PARTICLES

NEW DEVELOPMENTS IN PHYSICS

One cannot conclude this panoramic summary of the study of the nature of matter without stopping to consider the present situation in the field of sub-nuclear particle research. It is a situation which appears somewhat chaotic; and physicists think it will only be resolved as a result of some profound new development. But when will it be resolved?

The physicist Dyson, in an original and comprehensive investigation of the way in which the various new developments which have occurred in the history of physics came about, reaches an interesting conclusion. Although there are many exceptions a careful review of the history of physics nevertheless shows that as a general rule every great new development arises as the culmination of a gradual increase in knowledge which covers a period of about sixty years. Of these sixty years about thirty elapse between the discovery of a disconcerting phenomenon and the birth of a new idea which allows one to interpret it. The remaining thirty years run from the moment in which this new idea is put forward to the moment at which the most important consequences are produced from it. The first thirty years are a period of bitter struggle and of searching for a solution; the second thirty years are passed in systematic development and in the assimilation of strange and unaccustomed concepts.

Thus thirty years elapsed between Faraday's discovery of electromagnetic induction and Maxwell's theory which was introduced to explain it; and a second thirty years passed between Maxwell's theory and the demonstration by Herz of the existence of electromagnetic waves. Again, thirty years divided the discovery by Rutherford of the atomic nucleus and the first approximate understanding of its structure and of nuclear reactions; and there was a second period of thirty years between the quantum mechanics of Heisenberg and the new theory of super-conductivity by Bardeen.

Naturally there do not exist in science rigorous historical laws and one could not extrapolate the regularities which can be observed. But even if one should not accept the 'sixty years law' one should nevertheless recognize that the process of understanding which precedes and which follows every notable new development is slow and extended; and that the work which scientists, both great and undistinguished, must carry out in order to absorb and assimilate a new development after it has occurred is as long, arduous and important as that which drives them and carries them forward to new efforts along the road which leads to the birth of new developments.

What forecast can one then make about the future developments of the present situation in the physics of elementary particles? I will leave until later the reply to this question; this is not because I want to introduce into this chapter the suspense appropriate to a detective story but because I believe it is necessary first to describe the present situation which I defined above as somewhat chaotic.

WHAT DO WE MEAN BY 'ELEMENTARY PARTICLES'?

The phrase 'elementary particles' has been used in the past to indicate those particles which appear not to be made up of other smaller particles and which can be considered as the ultimate constituents of matter. In reality however the story of physics has shown, more than once, that the property of being 'elementary' which at a given period has been attributed to certain particles was due only to the limitations of our knowledge. Thus the physicists and chemists of the second half of the last century believed that the atom was the ultimate particle of matter, which could not be further decomposed; i.e. they thought that the atom was an elementary particle. But when it was discovered that the atom was in fact a complex system made up of a nucleus and of electrons the concept of 'elementary' was transferred to these smaller objects; it was said that the electron and the nucleus were elementary particles. Later when it was found that the atomic nucleus is in its turn made up of protons and neutrons the quality of 'elementary' was transferred yet again: it was said that the electron, the proton, and the neutron were elementary particles....

This shows that the adjective 'elementary' should be interpreted only as relative to our understanding; at every stage in the development of science one calls 'elementary' those particles which do not have structure or at least have a structure which cannot be described in terms of other smaller particles. This is why it is suitable to give to our particles the less exacting name of 'sub-nuclear particles'.

I have already said that in recent years the number of sub-nuclear particles which have been discovered is steadily growing. As Fermi pointed out, we are in a situation analogous to that of a physicist or a chemist who knew nothing of atomic structure and who, in starting to investigate the nature of chemical substances, discovered for example an atom of iron; he might think that this was an elementary particle. But if he then discovered an atom of oxygen and later one of sulphur and again one of silver ... inevitably his confidence that the atoms were elementary would be shaken; such a high number of particles would be in contradiction with the concept that the particles were elementary.

Physicists today are in a similar situation; there are about thirty particles which are called elementary (see Table III on page 286). We accept that it is hardly possible that all these particles are elementary. It is very likely that when this situation is cleared up the problem of nuclear forces will at the same time be resolved.

THE FIRST PARTICLES

Let us now review the history of development from the year 1930 on. It is a story in full evolution and development; research is carried on by an army of physicists, experimental and theoretical, and their results are published in exceedingly specialized articles in reviews throughout the world. To summarize the situation is exceedingly difficult. Not very long ago, the physicists Gell-Mann and Rosenbaum published an extremely clear picture of the theoretical ideas which guide physicists in their attempts to understand the material world;

TABLE III

	Particles	Spin
Ξ (XI) / Sigma / Lambda / Nucleon (proton, neutron)	Barions	$\frac{1}{2}$
K / Pion	Mesons	0
Muon / Electron / Neutrino	Leptons	$\frac{1}{2}$
Photon		1

it appears that these ideas are beginning to display a certain order in the chaos of sub-nuclear particles. This chapter will follow the steps of that review.

In about 1933 there were only four characters on the atomic scene: the electron, the proton, the neutron and the photon. The first three of these are the particles which make up atoms: the proton and the neutron of the atomic nucleus and the electron which revolves around them. The photons are the quanta of the electromagnetic field and form the first example of a type of particle (bosons) which are essentially different from the three preceding particles, electron, proton and neutron.

Between the photon and these other three particles there exists then another fundamental difference. The photon has zero rest mass; that is, it cannot exist at rest but always moves with the velocity of light. The electron, the proton and the neutron on the other hand have a rest mass; that of the electron is often taken as the unit of mass in sub-atomic physics. With this unit the mass of the photon is equal to 1836·1 and that of the neutron equal to 1838·6.

We already know well the value of the electric charge of these particles; the charge (negative) of the electron is often taken as the unit of electric charge; that of the proton has the same value but is positive; the neutron has no electric charge. Finally the photon, although it carries electromagnetic energy, does not have electric charge.

THE SPIN OF ELEMENTARY PARTICLES

When we were discussing atomic physics I drew attention to the so-called 'hypothesis of electron spin' which was introduced in 1925 by Uhlenbeck and Goudsmit in order to account for the optical and magnetic properties of atoms; electron spin later appeared in Dirac's theory as a necessary result of the introduction of relativity into the quantum theory; it thus ceased to be a mere *ad hoc* postulate.

Thus the electron has a *spin* (that is, a mechanical angular momentum) and a well determined magnetic moment; these properties and their numerical values are essential properties of the electron.

The possession of a spin is not however a characteristic only of electrons; every particle has a spin whose value constitutes one of its characteristic properties.

Thus every particle has always the same well-determined spin (that is, in the final analysis, it has the same velocity of rotation. The spin is thus an intrinsic property of a particle just as its mass and its electric charge.

We should note an extremely important fact. If one takes as the unit for measuring the spin of an elementary particle the quantity $\frac{h}{2\pi}$ (where π is the number 3·14 and h is Planck's constant) one finds that the spin of elementary particles can have only one of the following values:

$$0, \tfrac{1}{2}, 1, 1\tfrac{1}{2}, \ldots$$

that is, only an integral value or half-integral value.

We already know that the electron has a spin of $\tfrac{1}{2}$. It is found that protons and neutrons also have, like electrons, spins equal to one-half. The photon on the other hand has a spin equal to one.

I said finally that the spin cannot have any arbitrary direction but only certain well-determined directions (Fig. 40). In particular if the spin is equal to one half it can only be disposed parallel or anti-parallel with respect to an external magnetic field; this, as we have seen occurs for the electron which has a spin equal to one half.

These results, which have been confirmed by experiment, are yet another demonstration of that discontinuity in nature which appears in all atomic phenomena and of which I have spoken at length in Chapter Three. We saw there that an atom can exist only in a certain number of quantum states, that is that its energy can have only a discreet number of values; and we have now seen that the spin of a particle can take only one value and that it can orientate itself only in certain positions with respect to an external magnetic field (Fig. 40). These are all quantum effects due as usual to Planck's constant h.

STATISTICS

Every particle is then distinguished by its mass, by its electric charge and by its spin. It has, however, an additional property which is related to its spin: this property is called its *statistics*.

In the chapter on the atom I showed that electrons obey Pauli's exclusion principle. This requires that each quantum state of an atom may be occupied only by one electron; that is, in an atom there may exist only one electron revolving about the nucleus in a given orbit (allowed) and which has its spin orientation in a given direction (amongst those which are allowed).

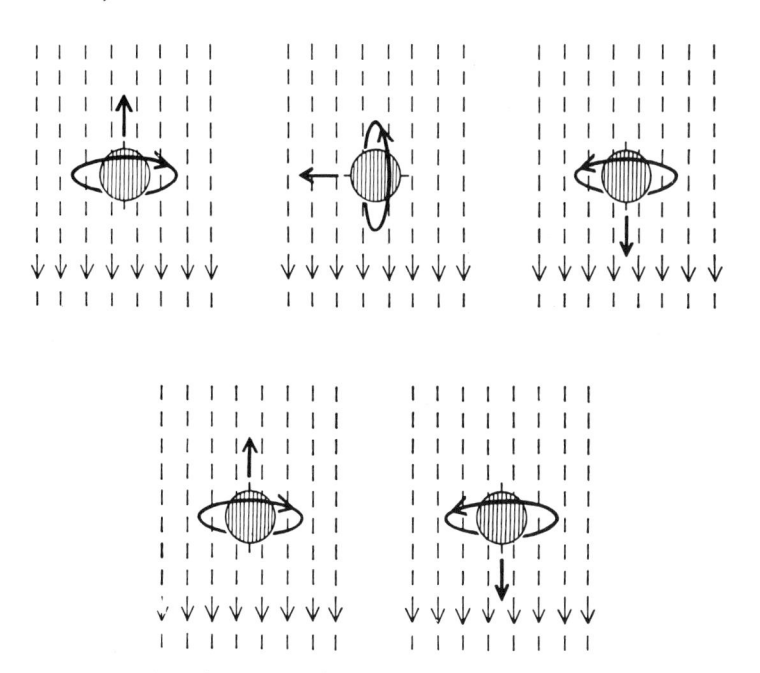

FIG. 40. Orientation of the axis of a particle in an external magnetic field.

It is, however, not only electrons which are required to obey Pauli's exclusion principle. *All particles which have spins of one-half obey Pauli's exclusion principle*; that is, given various different particles all with spin one-half, in a given quantum state there can exist only one of each type. Thus, electrons, protons and neutrons which all have a spin of one-half, all obey the exclusion principle; they are said to follow Fermi–Dirac statistics and are called *fermions*.

On the other hand, all particles which have zero or integral spin (like the photon which has spin one) do *not* obey the exclusion principle; they are said to follow Bose–Einstein statistics and are called bosons.[1]

These considerations on the spin and on the statistics of particles, which can only appear convincing as a result of a deep understanding of the subject, are essential for an attempt to make a panoramic picture of the different particles and to try to find in this picture a certain pattern.

We must always bear in mind the importance of this fundamental distinction between bosons and fermions; it concerns two types of particles which have essentially different properties.

On page 286 we reproduce a table which shows how the number of sub-nuclear particles has grown during the last thirty years.

At this point in the story we are in column one. There are four sub-nuclear particles; two of these are heavy (the proton, p^+ and the neutron n^0); one is light (the electron e^-)[2] and one is of a special type (the photon which is indicated with the Greek letter γ (gamma). The significance of the arrangement in the table of the horizontal lines, etc. will become clear only later.

INTERACTIONS BETWEEN PARTICLES

As we now know these four elementary particles are charac-

---

[1] The so-called Bose statistics were formulated by Bose in 1924 in order to account for Planck's radiation laws; these laws could not be obtained by applying to a photon gas the classical statistics of Boltzmann. Einstein then applied Bose statistics to an ordinary monochromic gas. The result of the two statistics differed considerably only when the temperature of the gas considered is extremely low or when the pressure (and therefore the density of the gas) is extremely high.

Shortly after Bose had formulated his statistics Fermi proposed a different type; he was guided by the exclusion principle which had been put forward by Pauli for atomic electrons and which was held by Fermi to be much more generally valid. Fermi's statistics also lead to the same results as classical statistics for normal conditions of temperature, density and pressure.

Wave mechanical treatment later showed that both of these statistics are equally valid and that they are mutually exclusive in the sense that it is found that if in a particular case one type of particle obeys one of these statistics it will always obey them. The choice between the two statistics is determined for the different particles by the value of their spin; if this is equal to a half they follow Fermi–Dirac statistics; if it is equal to nought or a one they follow the statistics of Bose–Einstein.

[2] Generic names for the heavy and the light particles are respectively, *barions* and *leptons*; these terms are derived from the Greek words baros = heavy and leptos = light.

terized by the values of their mass, their electric charge, their spin and magnetic moment, in addition to other qualities which I shall try to define later.

Amongst the properties which characterize a particle one should consider also the forces with which it interacts with other particles. There thus arises the problem of examining the laws of interaction between elementary particles. Fortunately, although the number of particles is about thirty the type of interactions are, as we shall see, very few: there are only four or three depending on whether one includes or does not include gravitational interaction.

The first type of interaction to be studied is the interaction between the electron and the photon, which is described by the so-called quantum electrodynamics. According to this theory the electromagnetic field which is produced in the space which surrounds a charged particle (for example an electron) should be described as a continuous succession of processes of emission and absorption of photons. This elegant theory which is based on Maxwell's classical theory of electromagnetism has been confirmed by innumerable experiments, so that it forms the model on which we try to build the description of other types of interaction.

This process in which electrons continually and rapidly emit and absorb photons is an example of what has come to be called a virtual process; it is characteristic of quantum theory. According to classical electrodynamics an electron cannot emit electromagnetic radiation (i.e. photons) and at the same time preserve unaltered its energy. Such a process is forbidden by the law of conservation of energy; in fact the total energy in the system would abruptly increase as the act of emission. In quantum electrodynamics on the other hand this process becomes possible because the emission of the photon is immediately followed by its absorption; in this way the increase in the total energy of the system is very, very transitory so that it cannot be measured. And when we remember that according to the quantum theory the laws of physics can be applied only to those quantities which (at least in principle) can be measured we realize that the conservation of energy is not in fact violated.

This virtual process of emission and absorption of photons can become real only when the conditions are such as to make possible the conservation of the total energy of the system; thus, if, for example, a moving electron is decelerated one or more real photons are emitted and the energy they carry is equivalent to the loss of energy of the electron.

By means of these four particles (electrons, protons, photons and neutrons) the theory was able to explain all the properties of atoms and to account for the charge of the atomic nuclei and approximately for their masses. 'But', in the words of Gell-Mann and Rosenbaum, 'nothing in the theory explains why nature chooses just these particles as its bricks for building; though given that they had been chosen, they were almost all that was needed.'

Almost all, but obviously not all.

THE ANTI-PARTICLES

Dirac had, as a result of his theory of the electron, deduced the existence of the positron (that is, of an electron with positive electric charge, ϵ^+); these particles were, as we have already seen, observed for the first time by Anderson. A pair of positive and negative electrons is produced by the materialization of a gamma quantum; and vice versa when positive and negative electrons collide together they annihilate themselves and disappear with the emission of two or more gamma quanta.

As the theory of Dirac refers to particles with spin one-half (i.e. to fermions) and as the proton and the neutron have, like the electron, spin one-half it was natural to think that for each of these particles there could also exist corresponding anti-particles.

The fact that even the neutron has its anti-particle shows that in general the concept of anti-particles is more complicated than that which one normally believed. In fact, as the neutron is electrically neutral, how could an anti-neutron exist, if the characteristic of an anti-particle were only to have electric charge of sign opposite to the electric charge of the corresponding particle? I will not dwell here on this question; for I do not want to interrupt the logical course which we are laboriously following in the attempt to find some pattern in the complexity of phenomena concerning the elementary particles which are known today. I will return to it later. For now, it is sufficient

Sub-Nuclear Particles

to know that the proton and also the neutron have their anti-particles.

Although some processes which could be attributed to anti-protons were observed in the cosmic radiation the definite proof of their existence was obtained only in 1955 by Chamberlain, Segré, Weigand and Ypsilantis, using the bevatron of the radiation laboratory at Berkeley (California); one might say that this machine was built specifically for this purpose. And shortly afterwards, again with this machine, anti-neutrons were also produced and observed.

Finally, from a mathematical point of view, the photon also has its anti-particle; but in this case particle and anti-particle are indistinguishable. We can say that the photon is the anti-particle of itself.

And so the sub-nuclear particles have now become seven. We are now at column two of our table in which the particles are indicated by white circles and the anti-particles by shaded circles. The heavy particles, that is the baryons (protons, neutrons and their anti-particles) lie in the first row and the light particles, that is the electrons (electrons and positrons) in the third row.

THE PION AND THE NEUTRINO

Other particles were very soon added to these seven. We already know them; they are the pion (positive, negative and neutral) and the neutrino.

The existence of pions or π mesons had been predicted by the Japanese physicist Yukawa (as we already know) in his attempt to explain the forces which hold protons and neutrons bound together in a nucleus. As the electromagnetic forces had been explained with such success by means of photons (i.e. of quanta of the electromagnetic field) Yukawa proposed a similar explanation for the nuclear forces. Just as an electron continuously and rapidly emits and absorbs photons (in that virtual process to which I have recently referred) so also nucleons continuously and rapidly emit and absorb a quantum of the nuclear field; Yukawa called this quantum the meson and today it is called the pi meson, or pion; it has a mass equal to about 270 times the mass of the electron and exists in three forms:

positive (π^+), negative (π^-) and neutral (π^0). The negative pion is the anti-particle of the positive pion and vice versa; the neutral pion is, like the photon, the anti-particle of itself.

Naturally this process of continuous emission and absorption of pions by the nucleons is also a virtual process. Furthermore, in order to account for the great strength of the field of nuclear forces we must assume that the emission of pions occurs so frequently that there are normally not one but several pions which exist at the same time outside the nucleon; and so we think today of protons and neutrons as if they consisted of a type of kernel surrounded by a pulsating crowd of pions. And finally as in the case of photons, if a sufficiently large amount of energy is furnished from outside (and it is found that this energy must be greater than 135 Mev), the virtual pions are materialized into real particles.

In column three of the table we have therefore added the pions (π); as they have a mass intermediate between that of the heavy particles (nucleons) and that of the light particles (electrons and positrons) they are assigned to this intermediate layer.

We will now go to column four. To the ten particles we add the neutrino. The existence of this particle was, as we know, proposed by Pauli in order to account for the behaviour of the neutron which, when unbound, decays in 12·3 minutes into a proton and an electron with the apparent disappearance of a small amount of mass; this is carried away in the form of energy by a neutrino (ν) a particle with zero rest mass which can only with great difficulty be observed.[1] Fermi, starting from this hypothesis of Pauli's, constructed a complete theory of beta disintegration of radioactive substances which accounted for the observed distribution of energy with which the electrons were emitted. Again in this theory the fundamental process was a virtual process in which a neutron continuously emits and absorbs an electron and a neutrino. In this case, however, the process can become real without energy being supplied from outside; this energy in beta decay is supplied by the mass which disappears.

The neutrino has its anti-particle; indeed, Fermi's neutrino was in reality an anti-neutrino.

---

[1] It has recently been discovered that this neutrino is not identical with that found in beta disintegration. We shall return to this point on p. 309.

Sub-Nuclear Particles

TWELVE PARTICLES

We are now at the end of the first part of the story. We have already collected a fair number of particles; but it appears that all are significant in the sense that each of them finds a place in the theoretical description of the phenomena which we have described.

Let us for a moment consider our separate particles as a whole and see if they display some pattern. First of all they can be divided into four well-defined groups (the horizontal layers in the table). (1) heavy particles (baryons: protons, neutrons and their (anti-particles); (2) particles of intermediate weight (mesons); (3) light particles (leptons: electrons, neutrinos and their anti-particles); (4) one type on its own, the photon.

These four groups are connected amongst themselves by three different and fundamental processes. These are: Yukawa's process (nuclear interaction) which relates the heavy particles with the mesons; Dirac's process (electromagnetic interaction) which relates electrically charged particles with photons; Fermi's process (beta decay) which relates the heavy particles with the light particles.

Furthermore, both the heavy particles and the light particles (which are the 'ordinary' constituents of matter) have spin one-half and are fermions (that is, they obey Pauli's principle): on the other hand the mesons and the photon (which are field quanta: of the nuclear field in the case of the mesons and of the electromagnetic field in the case of the photons); these quanta have integral spin (respectively zero and one) and therefore they are bosons.

Again all of these twelve particles obey the most general laws of physics: the conservation of energy, the conservation of momentum (linear and angular) and the conservation of electric charge.

Finally the reactions between these particles have two characteristics: (1) they are reversible; (2) the emission of a particle is linked with the absorption of the corresponding anti-particle; if we know the probability for the emission of a particle we can calculate the probability for the absorption of its anti-particle.

All of this supplies a certain number of rules with which we can resolve problems in the physics of the particles. For example, how can we explain the experimental result that a neutral pion decays (in an exceedingly short period of time, about 10^{-16} seconds) into two photons? That is

$$\pi^0 \to \gamma + \gamma$$

According to Yukawa's reaction, a proton emits a pion (virtual); that is

$$p \xrightarrow{v} p + \pi^0$$

As the reactions are reversible we can say that a proton (virtually) absorbs a pion and yields a proton:

$$p + \pi^0 \xrightarrow{v} p$$

And since in the equation which describes those reactions it is always possible to transfer a particle from the left-hand side to the right, providing that we substitute for it anti-particle, we obtain:

$$\pi^0 \xrightarrow{v} p + \bar{p}$$

where p̄ indicates an anti-proton.

But we know that the proton and the anti-proton annihilate each other giving place to two photons:

$$p + \bar{p} \to \gamma + \gamma$$

and therefore

$$\pi^0 \to \gamma + \gamma$$

which is just that which is observed in experiment. Naturally this argument is not as crude as may appear here but is supported by a rigorous calculation of the various probabilities.

But things did not always go as smoothly as this. This is what happened.

THE MUON

If we apply the rules which I have just explained to the decay of an electrically charged pion we find that the positive pion should decay into a positron and a neutrino. But this is very far indeed from what is observed to happen; it is found that it decays into a neutrino and into a new particle, which now enters for the first time on to our scene: the mu meson or muon (μ).

Here, in the words of Gell-Mann and Rosenbaum, nature shows herself to be particularly perverse: here she produces a particle for which there is no theoretical explanation and which is of no use at all. Furthermore the situation appeared at first even more complicated because of a very unfortunate accident of history . . .: the muon was discovered before the pion and it was believed (as I have already noted in the chapter on the nucleus) that the muon was the meson which had been predicted by Yukawa as the quantum of the nuclear field. But a fundamental experiment carried out at Rome between 1941 and 1945 by the physicists Conversi, Pancini and Piccioni showed, as I have already said, that all of its properties were completely different from those which it should have had in this case; before the pi meson was discovered (which is in fact the quantum of the nuclear field) the mu meson was even more disturbing than it is today.

The muon can have a positive or negative electric charge. It has a mass which is about 205 times greater than the mass of the electron and has spin of one-half (that is, it is a fermion). It has a half life of about one millionth of a second and decays into an electron (if it is a negative muon) a neutrino and an anti-neutrino; if it is positive it decays into a positron, a neutrino and an anti-neutrino; each muon of a given sign is the anti-particle of the muon with the opposite sign.

Let us then find a place for the muon in our table. As it is a meson we would at first think of putting it in the second layer, in which I have already put the pi meson. However, we are going to put it in the layer belonging to the light particles (electrons); we do this because the muon has in common with the electron the characteristic of being a light fermion.

The muon is perhaps the most mysterious particle in physics. A well-known physicist has written, 'it has not been possible to find any good reason for the existence of the muon, nor even to explain why it should have such a large mass.'

STRONG INTERACTIONS AND WEAK INTERACTIONS

Before going further we should consider for a moment the three fundamental processes according to which the various particles interact amongst themselves: the nuclear interaction of Yukawa, the electromagnetic interaction of Dirac, and Fermi's process of beta decay.

These three processes differ tremendously in their strength. Yukawa's process, which accounts for the exceedingly powerful source which binds the nucleons together in the atomic nucleus, is said to be a strong interaction. The electromagnetic forces (which come into play in Dirac's process) are about 140 times weaker than the nuclear forces. Finally, Fermi's interaction is very much weaker than the other two; it is one hundred billion times (10^{-14}) weaker than the strong interactions; it is referred to as a weak interaction. The terms strong interaction and weak interaction should be interpreted with respect to the electromagnetic interaction which forms a type of standard for comparison.

The stronger a process is, the greater is the probability that it takes place in a given interval of time; the greater that is the speed with which it takes place. Strong interactions therefore are the fastest that are possible: for example the emission and absorption of a pion takes place in about 10^{-23} seconds (ten billionths of a billionth of a second). In order to get some idea of the shortness of this time we should note that light, although it moves with a velocity of 300,000 kilometres a second, would travel in this interval of time only a distance equal to the diameter of an atomic nucleus (3×10^{-13} centimetres).

Electromagnetic processes are about 140 times slower.

It is found that all weak interactions have a type of intrinsic velocity which is about a thousandth millionth (10^{-9}) of a second (the actual velocity in a given case depends also on the energy available).

Thus for strong interactions the scale of time is about 10^{-23} seconds and for weak interactions it is about 10^{-9} seconds. This shows that the weak interactions are extremely slow compared to the strong interactions. In fact, the ratio between 10^{-23} and 10^{-9} is such that if we imagine extending the scale of time for the strong interactions to one second (instead of 10^{-23} seconds) the scale of time for weak interactions would become about a million years!

NEW STRANGE PARTICLES

The muon can be regarded as a hint by nature to the physicists indicating that they have not yet discovered all her hidden secrets. But in about 1950 she added to it a severe blow in the form of a series of new particles which were not only completely unexpected but which could not in any way be explained by the theories which had been so laboriously constructed.

These were discovered by the Englishmen, Rochester and Butler, in the showers which are formed when very energetic cosmic rays collide with a layer of material, such as a lead plate placed in a Wilson chamber. In studying these showers Rochester and Butler observed strange V-shaped tracks which could not have been produced in any process then known. They concluded that some unknown neutral particle had been produced in the lead plate and had decayed into two charged particles (the neutral particle naturally left no trace in the Wilson chamber).

From the study of a certain number of these events it was discovered that there are at least two types of neutral particles, one which, called lambda (Λ), decays into a proton and a negative pion; the other, the K^0, decays into a positive pion and a negative pion.

When they had recovered from the first shock physicists began to establish the properties of these particles. The lambda is a fermion (probably with spin one-half) and has a mass equal to about 2,181 times the mass of the electron; it is created in a Yukawa type interaction, as can be deduced from the frequency with which it is produced. In the table we will place this lambda particle with the heavy particles.

The K^0 particle is a boson (spin zero) and has a mass equal to 965 times the mass of the electron; this is also frequently produced and therefore arises from strong interactions. We will put it with the other bosons in the group of mesons where the pions are already established.

But in the study of the decay of these new particles, lambda and K^0, an extremely strange fact was discovered. These particles are made in strong interactions, that is, in a time of about 10^{-23} seconds; by the principle of reversibility they should therefore decay in a time of about the same magnitude. But this is not what happens. Although there is ample energy available for the decay this nevertheless takes place in a time between 10^{-8} and 10^{-10} seconds, that is in a time which is characteristic of the weak interactions. They live for an interval of time which is a hundred million times greater than that for which they ought to live.

Here we have an exceedingly strange phenomenon; it is so strange that we should not be surprised that these particles were called strange particles.

ASSOCIATED PRODUCTION OF STRANGE PARTICLES

For about two years physicists contemplated with bewilderment the great strangeness of these strange particles (Λ and K). Finally Pais put forward an idea which was able to explain why these particles are produced in a strong interaction process and decay by means of a weak interaction.

Since one can show theoretically that *one* strange particle which is produced in a strong interaction should also decay by a strong interaction and since this does not take place Pais suggested that for some reason strong interactions never give rise to a single strange particle (a single Λ or a single K), but always to two or more of them; for example, to a lambda particle and a K particle. But these two particles cannot then die through a strong interaction; they do not have enough energy to do so; indeed, their decaying products would have a combined mass greater than the mass available.[1]

The two (or more) strange particles, being unable to die immediately in a strong interaction, separate from each other as soon as they are born and continue to live; finally after a fairly long time (about 10^{-10} seconds) they die in that weak process which is much less likely and therefore much less frequent.

This hypothesis of the *associated production* has been sustained by experiment in the sense that there has never been

---

[1] For example, let us suppose that a lamda particle and a K-particle are produced in the collision between a proton and a negative pion
$$\pi^- + p \to \Lambda^0 + K^0$$
The inverse reaction would be
$$\Lambda^0 + K^0 \to \pi^- + p^+;$$
and since the absorption of a k^0 is equivalent to the emission of the corresponding anti-particle
$$\Lambda^0 \to \pi^- + p^+ + \overline{K}^0;$$
it is clear that the sum of the masses of the three daughter particles would be greater than the mass of the mother particle.

observed any process in which a single strange particle has been produced.

This concept has naturally immediately given rise to a new question; why are strong interactions which involve the production of a single strange particle forbidden? What is it that forbids it?

Physicists know that there are many things in nature which are forbidden. It is for example forbidden for a moving billiard ball, when it collides with a stationary ball, to come to a halt without setting the stationary ball in motion; this is forbidden by the law of conservation of energy. It is forbidden in a chemical reaction that the total mass of the reaction products should be greater than the mass of the substances which have taken part in the reaction; this is forbidden by the law of conservation of mass (which is valid in the macroscopic field). If a certain process cannot take place this is often due to the fact that it infringes some conservation law. The validity in nature of these conservation laws (which arise from symmetry laws) is a source of great encouragement to physicists; as the physicist Yang said in the speech which he made on receiving the award of the Nobel Prize for physics in 1957, 'the physicist learns in this way to hope that there is in nature a system which he can hope to understand.'

Might not then the associated production of strange particles (in strong interactions) be imposed by some conservation law? Which one?

THE LAW OF CONSERVATION OF 'STRANGENESS'

This law was discovered independently by Gell-Mann and Nishijima and is the law which requires conservation of a new physical entity especially introduced in this field; it is called *'strangeness'*.

One might be able to explain the nature of this new physical quantity; but to do this it would be necessary to introduce yet another new concept the concept of *isotopic spin* and to discuss the values which it assumes for the different particles (old and new). I think that this would lead to a confusion of ideas; for isotopic spin is a rather abstract concept which was introduced by Heisenberg in order to characterize the proton and neutron as two states of a single entity—the nucleon; one can say approximately that these two states differ only in the value of their electric charge. This concept was then extended to all the other particles.

I will restrict myself here to saying that the 'strangeness' of a particle is a number which can have only integral values (positive and negative) and which is, crudely, a measure of the amount by which the isotopic spin of a particle differs from the value which it should have. The non-strange particles therefore (protons, neutrons, anti-protons, anti-neutrons, pions, etc.) have strangeness zero. The lambda has strangeness minus 1 and the anti-lambda plus 1; the K plus 1 and the anti-K minus 1. Strangeness numbers different from zero are attached also to the other strange particles which were later discovered: the sigma particle (Σ) and the xi particle (Ξ).

Once these new quantities have been introduced one can show that they must be conserved in processes in which strong interactions and electromagnetic interactions are involved. In every reaction of one or the other of these two types the total strangeness[1] of the particles which enter into the reaction should be equal to the total strangeness of the products of the reaction. In weak interactions on the other hand strangeness is not conserved.

It is this law of the conservation of strangeness in strong interactions which requires the associated production of strange particles. These strange particles are in fact produced in collisions between ordinary particles; these have strangeness zero and therefore the total strangeness of the products must also be zero. In our example of the production of a lambda and a K particle in the collision between a proton and a pion the process can take place because the lambda has strangeness minus 1 and the k-particle plus 1.

Associated production then explains why strange particles do not decay by means of a strong interaction. One can show further that the law of conservation of strangeness does not even allow them to decay by means of an electromagnetic interaction. They can therefore only decay into ordinary particles in a time of the order of that which is characteristic of

---

[1] Defined as the algebraic sum of the strangenesses of the single particles.

weak interactions for which as I have said the principle of the conservation of strangeness is not valid.

THE PRINCIPLE OF PARITY

As the physicist Morrison has pointed out when the philosopher Leibnitz put forward the principle 'two states which are indistinguishable from each other are the same state' he established one of the most solid pillars of modern physics which sustains not only the theory of relativity but also the conservation laws on which are based our understanding of nature.

One of these laws is the law of *conservation of parity*; it deals with the indistinguishability of left and of right and crudely can be defined in this way: there is no absolute distinction between a real object (or an event) and its mirror image. If I look at the development of an event in a mirror, for example at a man who is writing on a blackboard, I realize that I am looking at a mirror image from the fact that the writing is reversed and that if the man turned to face me he would use his left hand instead of his right hand. But it could easily be that the man is left-handed and that he, like Leonardo da Vinci, wrote from the right towards the left. How can I with this doubt in mind, succeed in knowing whether what I see is a real event or its mirror image? No *intrinsic* property of what I see allows me to reply to this question.

From this principle of the indistinguishability of left and right it follows that one can always have an exact mirror image of any event whatsoever. Somebody might object that the heart of the man is on his left side or that the spiral of a shell always turns in the same sense. But there might be a world in which the 'right' was changed with the 'left'; in this world everything could carry on just as in our own world.

We reach then the idea that it is not possible to find any *intrinsic* difference which allows us to distinguish the physical phenomena which occur in the real world from those which would occur in a looking-glass world. Just as Dirac had shown (theoretically) that for every particle there should exist the corresponding anti-particle so the law of mirror-reflection affirms that if a particle exists there should also exist the particle obtained by reflecting it in a mirror; and that if a reaction can take place then the corresponding mirror image of the reaction should also be possible.

A consequence of this 'mirror invariance' is the principle of *conservation of parity* just as (though I will not dwell on this point) the invariance of physical laws with respect to translations in time[1] implies the classical principle of the conservation of energy.

But although it is easy to give a definition of energy it is not possible to give an intuitive definition of *parity*. Roughly one can say that parity is a property of a function which in quantum mechanics represents the position of a particle in space. If when we change the sign of one of the three spatial variables of this function (this is equivalent to reflecting the system in a mirror) the function itself changes sign we say that it has 'odd' parity; if on the other hand the function is unchanged we say that it has 'even' parity.

Therefore a system (for example a particle) has either even or odd parity.

Until a short time ago theory and experiment showed that in an isolated system the parity never changed; if it was even it always remained even; if it was odd it always remained odd. Theory and experiment agreed on the validity of the principle of conservation of parity.

The principle was regarded as self-evident until the summer of 1956 when two young Chinese physicists, Yang and Lee, working at Columbia University tried to resolve an apparent paradox which appeared in the investigation of strange particles. They suggested, with amazing temerity, that there might be an exception to the principle of conservation of parity; as the paradox appeared in the field of weak interactions they considered that there was no *a priori* reason why the principle of conservation of parity, which had been shown to be valid in other fields (electromagnetic interactions . . .), should also be valid in the case of weak interactions for which the characteristic time is, as we know, one hundred thousand million times greater than that of electromagnetic reaction.

They proposed an experiment which could be used to test

---

[1] That is the fact that there is no absolute time.

whether in the case of weak interaction left and right could be distinguished. The experiment was based on the following considerations. If a particle which revolves about itself (i.e. a particle which has a spin) emits another particle in the direction of its axis (Fig. 41) its mirror image could not be distinguished from it provided that the secondary particles are emitted with equal intensities in the two opposing directions: in the example

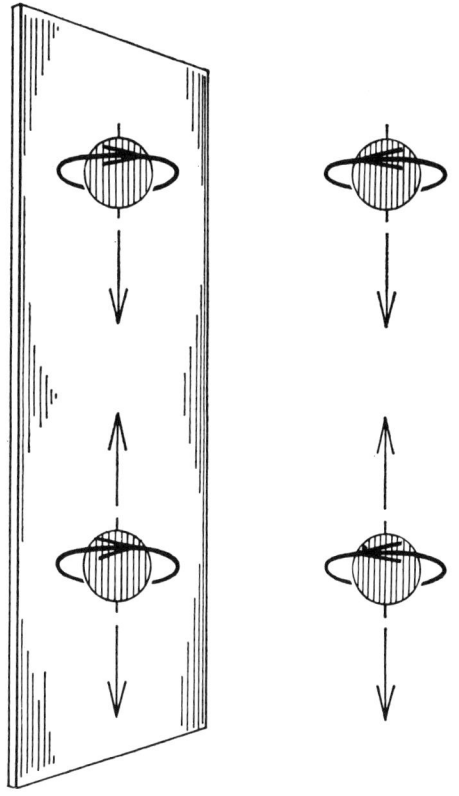

Fig. 41.

of the figure (in which the axis is vertical) upwards or downwards. But if the particles are emitted with greater intensity in one direction as opposed to the other (for example with greater intensity upwards) then the mirror image could be distinguished

306   *The Nature of Matter*

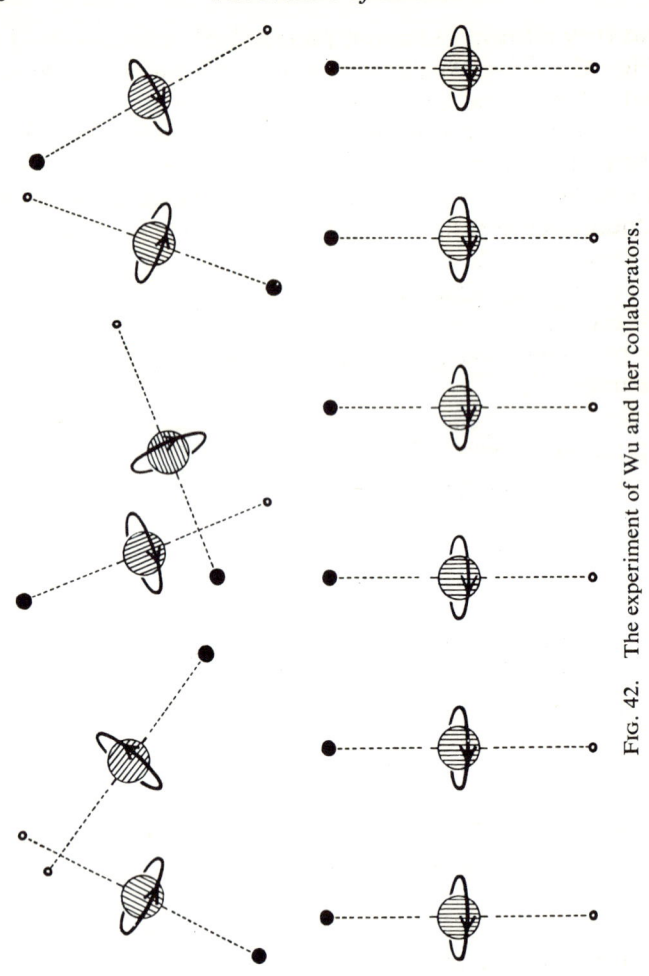

Fig. 42. The experiment of Wu and her collaborators.

from the real event. In fact in the case of the figure whilst in the real event the particle revolves towards the right of a person facing in the direction of the emitted particle in the mirror image the rotation would occur towards the left.

Yang and Lee proposed that radioactive nuclei of cobalt with atomic weight 60 should be used for the experiment; these nuclei decay emitting an electron and a neutrino (Fig. 42 on the left). They proposed that by the cooling and by the application of a magnetic field the cobalt nuclei should be aligned in such a way that they all revolved about themselves in the same direction (Fig. 42 on the right); in these conditions it should be observed whether the electrons were emitted with equal intensity in the two directions along the axis or whether on the other hand they were emitted with greater intensity in one direction rather than the other. In this second case there would be no doubt that in beta disintegration (which is a weak interaction) there exists a distinction between left and right, i.e. that it would be possible in this field to distinguish between a physical event and its mirror image. There would be no doubt that the principle of the conservation of parity would be invalid in weak interaction

Very few physicists at that time doubted the general validity of the principle of conservation of parity. On January 17, 1957 (as the physicist Salam has recorded) Pauli wrote to his friend Weisskopf: '... I do *not* agree that God is a "weak left-hander" and I am ready to bet a very large amount that the results of the experiment will be symmetrical.'

Two days later the experiment was carried out after six months of preparation by the Chinese physicist Mrs Wu and her American collaborators at the Bureau of Standards in Washington. It lasted for sixteen minutes and showed with no possibility of doubt that the nuclei of cobalt-60 when aligned in a magnetic field emits electrons with greater intensity in the direction opposite to the direction of the magnetic field.

The absolute invariance of the principle of parity was dead; although valid in other fields it is *not* valid in the kingdom of weak interaction.

On January 27, 1957, Pauli wrote; '... now after the first shock I am beginning to recover ... I am not shaken so much because the Lord prefers the left hand but because he appears left-right symmetrical when he expresses himself strongly.'

All of this furnishes physicists with a serious lesson. It may be that the fundamental conservation principles which are becoming steadily verified are not universal but are limited to the particular fields in which they have been shown to be valid. In the realm of weak interactions neither the principle of conservation of parity nor the principle of conservation of strangeness is valid; on the other hand it seems that in weak interactions the principle of conservation of energy is still valid but this should not lead us to claim that this principle is universally valid: it may be that tomorrow it will be found that for interactions still weaker than weak interactions (that is for those which involve the force of gravity) the principle of conservation of energy is no longer valid. Then the cosmological theory upheld by some astronomers would not appear too fantastic; these astronomers think that new matter is continuously being created in the expanding universe.

The research, the theories and the discoveries made in the field of elementary particles would then have incalculable consequences on our ideas on the constitution of the entire universe; micro-physics and cosmology will certainly be closely linked.

As the physicist Morrison has said the discovery of the limited application of the principle of the conservation of parity is not a discouraging set-back but a favourable opportunity. And he adds: we have entered into and are now living in an exceedingly exciting period.

WEAK INTERACTIONS AND NEUTRINO PHYSICS

Today one of the most important and fascinating fields of research in physics, both theoretical and experimental, is the study of the large class of weak interaction processes. It has been found so far that in these processes two of the fundamental conservation laws are not valid; the weak interactions, as we have seen, violate both the law of conservation of strangeness and the law of conservation of parity. Is there some relation between these laws and their violation by the weak processes? We do not know: nature has certainly hidden in the weak processes one of her deepest secrets.

We have seen on page 191 that the neutrinos emitted in the beta disintegration of the radioactive nuclei which are created in a pile are able to produce nuclear reactions; but now that we know that in addition to the neutrinos there exists also antineutrinos we are able to be more precise about the reaction which was observed by Cowan and Reines; it was in fact produced by anti-neutrinos ($\bar{\nu}$) according to the scheme:

$$\bar{\nu} + p \rightarrow n + \epsilon^+$$

We can express this in words by saying that an anti-neutrino is absorbed by a proton which is changed into a neutron with the emission of a positive electron (that is, an anti-electron).

The success of these first experiments on the reactions produced by neutrinos have given rise to the desire to investigate in more detail and in different circumstances the processes produced by these particles which have zero charge, zero mass and various other very odd properties. In particular there has arisen the problem of observing some nuclear interactions produced by neutrinos of high energy, such as those emitted in the decay of the unstable particles produced with the large accelerating machines.

In the first months of 1962 a group of American physicists (Danby, Gaillard, Goulianos, Lederman, Misty, Schwarz and Steinberger) succeeded in finding evidence for the reaction

$$\bar{\nu} + p \to n + \mu^+$$

they made use of the anti-neutrinos produced in the decay of the positive pions which were generated in the 30 Gev accelerator of the Brookhaven National Laboratory. This reaction is *different* from that observed with the anti-neutrinos produced in a pile.

This is a discovery of the greatest importance which confirms a suggestion which had been put forward a few years ago by Pontecorvo: the anti-neutrinos emitted in beta disintegration processes are different (in that they produce different reactions) from the anti-neutrinos emitted in the decay of pions. There are therefore two different types of neutrinos which are indicated respectively with the symbols ν_ϵ and ν_μ; each of these has its corresponding anti-neutrino, $\bar{\nu}_\epsilon$ and $\bar{\nu}_\mu$.

This is a fundamental new aspect of nature which had previously escaped observation; it is not easy today to foresee all the consequences. What one can say is that we now have opened a new chapter in physics, *the physics of the neutrino* and that today we have hardly reached the end of the first page.

A quick glance at all the particles so far named (in the penultimate column of the table) shows the validity of the fundamental idea of the existence of four groups of particles (1—heavy particles or baryons, some of which are strange; 2—mesons, some of which are also strange; 3—light particles

TABLE IV

Particle	Mass in units of the electron mass	Mean life in seconds
1. Photon	0	stable
2. Neutrino	0	stable
3. Anti-neutrino	0	stable
4. Electron	1	stable
5. Positron	1	stable
6. Muon, negative	206·8	$2·2 \times 10^{-6}$
7. Muon, positive	206·8	$2·2 \times 10^{-6}$
8. Pion, negative	273	$2·55 \times 10^{-8}$
9. Pion, positive	273	$2·55 \times 10^{-8}$
10. Pion, neutral	264	2×10^{-16}
11. K, negative	966	$1·22 \times 10^{-8}$
12. K, positive	966	$1·22 \times 10^{-8}$
13. K, neutral 1	974	1×10^{-10}
14. K, neutral 2	974	6×10^{-8}
15. Proton	1,836	stable
16. Anti-proton	1,836	stable
17. Neutron	1,839	1×10^{3}
18. Anti-neutron	1,839	1×10^{3}
19. Lambda	2,183	$2·5 \times 10^{-10}$
20. Anti-lambda	2,183	$2·5 \times 10^{-10}$
21. Sigma, negative	2,341	$1·67 \times 10^{-10}$
22. Anti-sigma-negative	2,341	$1·67 \times 10^{-10}$
23. Sigma, positive	2,328	$0·83 \times 10^{-10}$
24. Anti-sigma-positive	2,328	$0·83 \times 10^{-10}$
25. Sigma, neutral	2,331·7	$< 10^{-11}$
26. Anti-sigma-neutral	2,331·7	$< 10^{-11}$
27. Xi, negative	2 580	$1·3 \times 10^{-10}$
28. Anti-xi, positive	2,580	$1·3 \times 10^{-10}$
29. Xi, neutral	2,566	$\sim 10^{-10}$
30. Anti-xi, neutral	2,566	$\sim 10^{-10}$

or leptons; 4—photons), and of three types of interactions (strong interactions, electromagnetic interactions, and weak interactions).

These have been discovered from the regularity in the properties of the sub-nuclear particles. We must now discover the reason for the regularities.

Mendeleev had discovered certain regularities in the properties of the elements; the reason for these regularities was discovered only after Pauli had put forward his exclusion principle.

In the study of the sub-nuclear particles we are today in the same situation as Mendeleev; physicists are searching for the laws which explain why these observed regularities exist.

Recently a physicist has written, 'looking through the table of the elementary particles one cannot fail to reflect on how privileged is the present generation which finds itself faced with such a fascinating problem. . . . I think that the present stage is only a stage of transition towards an internal harmony and a universal and profound symmetry.'

I cannot conclude this panoramic review without recording that there has recently been discovered yet another phenomenon characteristic of the particles which take part in the strong interactions, i.e. of the neutrons and the baryons. In many processes these particles are not produced independently of the others but are produced as if bound together in such a way as to make a single entity; this single entity would without doubt be indicated with the name of a particle if it did not have a mean life so short that it always decays before it escapes from the nucleus where it is produced. These entities are usually given the name of *resonances*. For example, there is a resonance indicated by ω^0 which is produced in the annihilation of a proton with an anti-proton; the ω^0 has a mean life of about 10^{-21} seconds and decays into three pions, one positive, one negative and one neutral. In addition to this there are other 'resonances' which decay into two or three pions, and others again which decay into K mesons and pions or into a pion and a hyperon.

However, today it is clear that the particles listed in Table IV can 'link' themselves together, though for exceedingly short periods of time, in order to make more complex entities which are extremely elusive; these properties constitute one of the most exciting and promising problems of the physics of elementary particles.

FUTURE QUESTIONS

Only when the laws which account for the observed regularities will have been discovered can we tackle other questions.

Are these sub-nuclear particles really all of the elementary particles? Why has nature chosen just these bricks for all that she builds? Why does the electric charge of the elementary particles always have one of the three values 0, -1, $+1$?

Have we now reached the time for the reply to that question which I left in suspense at the beginning of this chapter? 'What forecast can we make about the future development of the present situation in the physics of elementary particles?'

About thirty years have passed since Fermi formulated his first general description of the weak interactions which are responsible for the decay of a radioactive nucleus by the emission of a beta particle. In 1957 Lee and Yang discovered that an essential characteristic of these weak interactions is the violation of the law of conservation of parity; and the experiments which have followed have allowed us to resolve the problem posed by Fermi of giving a precise description of these processes. If the sixty-year law shows itself to be again valid we are at the middle of our task: another thirty years or so will be needed in order to make clear the full significance of weak interactions.

The physicist has before him two enormous tasks: to study and to perfect the mathematics of the present theories and to explore the wide fields of physical phenomena which have been left on one side by the existing theories, such as gravitation and cosmology. It is probable that a satisfactory theory of elementary particles will require a description of the conditions which existed at the beginning of our universe.

THE ANTI-PARTICLES AND THE STABILITY OF MATTER

Before concluding I will turn once again, as I promised, to the anti-particles.

The fact that even the neutron has its anti-particle (the anti-neutron) led us to the conclusion that an anti-particle cannot be characterized *only* as having electric charge of the same magnitude but of opposite sign as the charge of the corresponding particle; there must exist also another physical quantity which, like the electric charge, has the same value but opposite sign for a particle and for the corresponding anti-particle.

There is another reason for the existence of this new physical quantity. Matter has existed for thousands of millions of years: that is, the electrons, the protons and the neutrons which make it up are stable. This stability appears surprising; for since mass and energy are equivalent one might ask why, for example, an electron has remained in existence as such for thousands of millions of years and has not been spontaneously transformed into, for example, a photon and a neutrino; or why does a proton not change itself into a positron (which has the same charge but a much smaller mass) by leaving the residual mass in the form of energy. As usual there must be conservation principles which inhibit these transformations.

Let us start with the electron. It is the lightest particle to have an electric charge; thus by the principle of the conservation of electric charge (which can never be created or destroyed) it can never be transformed into particles which like the photon and the neutrino do not have electric charge. The only thing which an electron can do in order to disappear, that is to transform itself into energy, is to annihilate itself in a collision with a positron; this annihilation is not prohibited by the principle of conservation of electric charge because the positron has an electric charge which has the same values but opposite sign to the charge of the electron. The total charge of the system electron plus positron is zero before the annihilation and is zero afterwards; for the photons into which they are transformed do not have electric charge.

This is the reason why the electrons in matter are stable. Let us see now what happens for the protons.

We see immediately that the principle of conservation of electric charge is not adequate to account for the stability of a proton. It certainly would not forbid a proton to change itself, for example, into a positron (which has the same charge but a smaller mass) and into energy; the electric charge before and after the transformation would have the same value and the same sign (positive); and it would not prevent a proton annihilating itself in a collision with an electron which has electric charge of the same magnitude but an opposite sign. Since this does not happen we must think as usual that it is prevented by another conservation principle.

It is therefore thought that beyond the electric charge there exists another type of charge completely different from it which has been called the *baryonic charge*; this characterizes all the heavy particles (or baryons) and only the baryons because only for them is it different from zero. In addition it is thought that the proton is the lightest particle to have a baryonic charge different from zero and that the principle of conservation of baryonic charge is universally valid.

Thus just as the stability of the electron is due to the fact that it is the lightest particle whose electric charge is different from zero so the stability of the proton is due to the fact that it is the lightest particle whose baryonic charge is different to zero. A proton cannot transform itself into a positron and give up energy because in doing so it would infringe the principle of conservation of baryonic charge; in fact a proton has baryonic charge different from zero whilst the positron and the photon have baryonic charge zero. And a proton cannot annihilate itself in a collision with an electron with the transformation of all their mass into energy because in such a process the baryonic charge would not be conserved; it would be different from zero before the annihilation and be equal to zero after the annihilation.

This then is the reason why protons are stable.

And here we are finally at the neutrons which as we know are, when free, unstable (with a mean life of about twelve minutes) but which when they are bound to protons in a nucleus are normally stable; only in certain atomic nuclei of radioactive substances can a neutron spontaneously change itself into a proton emitting an electron and an anti-neutrino. This fact immediately tells us that the neutron should have a baryonic charge equal to that of the proton since both the electron and the neutrino have baryonic charge zero.

The anti-baryons on the other hand (the anti-protons, the anti-neutrons, etc.) always have a baryonic charge of opposite sign to that of the baryonic charge of the corresponding particle. In this way the law of conservation of baryonic number automatically requires that the production of anti-baryons can occur only when coupled with baryons.

Thus the introduction of this new physical quantity—the baryonic charge which like the electric charge obeys its conservation principles—allows one to explain the stability of the protons and of the neutrons whilst the principle of conservation of electric charge accounts for the stability of the electrons. The three particles, electron, proton and neutron, these three bricks with which the whole of matter is built can exist without undergoing transformation because of the validity of amongst other things the principle of conservation of electric charge and of baryonic charge.

But only these are stable (the neutron only when it is bound in the nucleus). There may exist—and in fact there do exist—other particles; but these are heavier than the proton (and thus than the electron) and, almost as soon as they have been born, are changed into protons and neutrons or electrons with the creation of lighter particles and lose their residual mass in the form of energy; and they change themselves into one or other of the stable particles according to the value and the sign of their electric charge and of their baryonic charge; according that is to what is allowed to them by the conservation principles.

STILL MORE PARTICLES AND ANTI-PARTICLES

And so today we believe that for every particle there is a corresponding anti-particle; if the particle has electric charge or baryonic charge zero the same occurs for its anti-particle; if the particle has electric charge and/or baryonic charge not zero the corresponding anti-particle has charges of the same value but of opposite sign.

Thus the proton has positive electric charge and positive baryonic charge; the anti-proton has electric charge of the same value but negative and it has negative baryonic charge. The electron has negative electric charge and zero baryonic charge; the positron has positive electric charge and zero baryonic charge. The neutron has no electric charge but has positive baryonic charge; the anti-neutron has zero electric charge and negative baryonic charge.

Before the existence of the anti-proton had been experimentally demonstrated some people thought that this idea of barionic charge, although subtle and suggestive, was in reality without physical significance. But the experimental discovery of the anti-proton and then of the anti-neutron showed how solidly based were the theoretical predictions; these predictions had been further confirmed by the experimental discovery of other anti-particles.

ANTI-MATTER

Protons, neutrons and electrons form the constituent parts of all atoms, that is of all matter. We can then think that there may exist an 'anti-matter' whose atoms—or rather whose anti-atoms—are made up of anti-protons, anti-neutrons and positrons; there might exist for example an atom of anti-hydrogen which would be made up of an anti-proton, around which would revolve a positron; and an atom of anti-helium, in which about a nucleus made up of two anti-neutrons and two anti-protons would revolve two positrons, etc.

On this earth of ours which is made up of matter, anti-matter cannot continue in existence; in fact, as we know, particles and anti-particles annihilate themselves as soon as they come in contact; all of their mass is transformed into energy. But does anti-matter exist in the universe? This question certainly has enormous importance for cosmologists. And its importance is not restricted to cosmology; it could also have enormous importance for physicists.

The theory of Dirac and the generalizations which have followed it are made in such a way that particles and anti-particles are treated equally: to every reaction in which particles are involved there corresponds a reaction in which the respective anti-particles are involved; and in these two reactions things proceed in an exactly analogous manner. In other words, there has been growing in physics a new symmetry principle: the symmetry between matter and anti-matter.

If symmetry in matter and anti-matter exists a consequence is that the anti-particles should in the case of weak interactions violate the parity principle just as in weak interactions particles violate this principle. If an experiment were to show that particles and anti-particles do not violate in the same manner the parity principle then in the weak interactions one would have an intrinsic distinction between matter and anti-matter—a distinction which today is held to be forbidden by the principle of symmetry between matter and anti-matter.

If this were so for the weak interactions then neither the principle of conservation of parity (that is of left-right symmetry) nor the symmetry of matter and anti-matter would be valid. There would then appear a possibility: what would happen if at the same time one changed left and right, matter and anti-matter? Would it perhaps happen that this 'combined symmetry' would be valid even in weak interaction? There is as yet no reply to this question: the experiments (very delicate and difficult) are now being made. If, as now appears likely, the answer will be affirmative it would establish an idea which is beginning to grow and which appears fantastic: that the complete universe might after all conserve that symmetry which for matter alone is lacking in the case of weak interaction. All that would be necessary is that in some part of the universe there should exist a quantity of anti-matter equal to the quantity of matter which exists; and that for this antimatter the parity principle were violated in the weak interactions in the direction opposite to that in which it is violated by matter.

Some astronomers are looking for this anti-matter which (to bring peace of mind to the physicists very shaken by these violations of symmetric principles) should exist in some part of the universe.

DOES ANTI-MATTER EXIST IN THE UNIVERSE?

The only way of proving the existence of anti-matter is by looking to see whether in some part of the universe there occurs annihilation of matter and anti-matter with the conversion of all the mass into energy. Now, according to Einstein's equation, when a proton and an anti-proton collide at a low velocity they annihilate each other and let loose an energy of about 1·88 thousand million electron volts. One should therefore search the universe for sources of energy of this order of magnitude.

If anti-matter should exist in the universe it must clearly be either inside our galaxy or outside it.

If anti-matter exists in our galaxy it must be either in the interstellar gas or in some of the stars in our own system of stars.

Let us start with interstellar space. This is known to be full of a very rare gas, consisting essentially of hydrogen; to be precise, there is one single atom of hydrogen in every cubic centimetre. If anti-matter should exist in this gas the antiprotons when they collide with protons would annihilate themselves and give up energy which would appear primarily in the form of neutrinos, of gamma rays and of very fast electrons and positrons. The probability that the neutrinos and the gamma rays would be absorbed by the protons of the very rare interstellar gas is exceedingly small. They would escape from the galaxy. The electrons and positrons on the other hand, which carry only about ten per cent of the energy set free in the annihilations, would remain imprisoned in the galaxy by the magnetic field; they would finally give up their energy to the interstellar gas (exciting and ionizing the atoms and annihilating themselves).

Now the energy of the interstellar gas is known; from this we can calculate an upper limit for the amount of anti-matter that it contains by making the most favourable hypothesis that all of the energy comes from the annihilation of anti-matter. A simple calculation leads to the conclusion that this upper limit is equal to a ten-millionth part of the amount of matter.

Given this very small percentage of anti-matter which could exist in interstellar space it is extremely unlikely that there exists in our galaxies stars made up only of anti-matter even if (in some unknown fashion) anti-matter were able to separate itself from matter and to condense into stars.

There could therefore be only a small quantity of anti-matter in the galaxy. What can one say about its existence beyond the galaxy?

If there were galaxies made up completely of anti-matter and if one of these came into collision with a galaxy made up of matter an enormous quantity of energy would be liberated which we might be able to observe. Well two places have so far been observed which are from this point of view extremely interesting.

There are indeed two galaxies (one called Messier 87 and the other Cygnus A) from which we receive exceptionally powerful radio signals; the signals are so powerful that it is difficult to account in terms of any known process for the enormous energy emitted by these radio sources. Some astrophysicists have therefore suggested that the galaxy Messier 87 (which is anomalously bright, which has a lateral strip which is even brighter, and which is a powerful source of radio waves) might be a galaxy which is capturing a large amount of anti-matter from an anti-galaxy. The other powerful radio source, Cygnus A, which is 270,000,000 light years away is really, as can be seen from photographs, made up of two galaxies in collision. If one supposes that anti-matter is present in each of these galaxies in the same percentage as that in which it can exist in our own galaxy (that is one part of anti-matter to ten million parts of matter) one can calculate that this radio source should emit about 10^{44} ergs per second of radio energy: and this is just what is observed. Is this just a coincidence?

We conclude then that in our galaxy anti-matter may exist but only in very small quantities compared with the amount of matter. Outside the galaxy there might exist in remote parts of the universe other galaxies made up completely of anti-matter.

There appeared not long ago an objective review of this subject by two well-known astro-physicists (Burbridge and Hoyle) who dwelt on the serious difficulties which would arise if one accepted the fascinating idea that anti-matter could exist in the universe.

Whatever theory one accepts of the origin of the universe symmetry arguments require that if anti-matter should exist it and matter should have been created in the same amount. And since matter exists today matter and anti-matter must have been separated immediately after their creation for otherwise they would have annihilated each other; after the separation they would have condensed into galaxies and stars.

Matter and anti-matter could be separated only if there existed between atoms and anti-atoms a repulsive gravitational force. That is a gravitational force opposite to the force of attraction which holds bound together the atoms of ordinary matter: in a single word, anti-gravity.

If one reflects that in accepting this idea of anti-gravity one would destroy the fundamental basis of the general theory of relativity one understands the uncertainty and the reluctance of the greater part of physicists. This uncertainty and this reluctance could be overcome only on the basis of fact supplied by experiment: if one projected parallel to the earth a beam of anti-protons and one discovered this beam moved away from the earth one would have the proof of the existence of anti-gravity. In theory this experiment is possible; however only physicists in the future will be able to give a reliable answer to this question. Today certain considerations put forward by the American physicist Schiff appear to lead to the conclusion that anti-gravity does not exist so that between protons and anti-protons the gravitational forces are attractive just as those between protons and protons.

However, in the meantime, some physicists have allowed their imagination free play. Thus M. Goldhaber thinks that there might be two separate worlds, one made of matter and the other of anti-matter. He thinks that the universe originated from one single 'particle' which he has called by the name of 'universon'.

This universon immediately divided into two particles, the 'cosmon' and the 'anti-cosmon' which separated one from another and then gave place, one to our cosmos and the other to an anti-cosmos, which is beyond the reach of our observation. If in some place and in some manner a very small amount of anti-matter was injected into the cosmos this would explain the anti-matter which might be dispersed in the interstellar space of our galaxy and this perhaps is the cause of the energy emitted from the galaxy Messier 87 and Cygnus A.

For now, these are only fantasies, although exceedingly interesting for the extraordinary width of the field they treat. Only the future can give the final word. But did not Democritus twenty-four centuries ago dream of the atoms and of their properties?

Finally, I should add that the opposition to this idea of two separate universes has shown some sign of diminishing since the discovery of Yang and Lee; the question has arisen whether there might not exist after all in the universe a general symmetry on a much wider scale; this might require the existence in the universe of equal quantities of matter and anti-matter; for these, the parity principle, in the case of weak interactions, would be violated in opposite directions.

Perhaps a complete understanding of the physics of the elementary particles will be possible only when observations will have allowed us to reply to the question raised by cosmologists.

CONCLUSION

With these very uncertain forecasts I will end my long journey which, starting from the observations of the various substances which we can see in the universe, has led us farther and farther into the intimate constitution of matter (from molecules to the atom, to the atomic nucleus and to sub-nuclear particles) and which has finally brought us to consider the possibility of the existence of anti-matter.

Innumerable problems have been resolved; many theories have been constructed; but here there are still new problems for whose solution it will probably be necessary to bring forward completely new ideas.

Physicists are continuously at work inspired by an invincible and eternally youthful spirit of adventure.

INDEX

d'Abro, 98, 105
accelerating machines, 196, 212–14; constant voltage, 215–17; variable voltage, 217–25
actinium, 53
d'Agostino, 203
alkalis, electron layers of, 85
alpha disintegration, 186–8
alpha particle; and accelerating machines, 213, 214, 224; and artificial radioactivity, 199, 203, 207–8; binding energy, 249, 250; bombardment of aluminium, 199; bombardment of beryllium, 164-5; bombardment of nitrogen, 162–3, 195, 197; collision with nucleus, 60–3, 149–50, 186; discovery of, 55–6, 59–60; fusion, 255; penetrating power, 166; photographic emulsion method, 151; radioactivity, 55–6, 59–60, 117–18, 131–2, 186–8, 208, 213; rest energy, 214; scintillation, 142; stellar energy, 265–6; transformation of lithium into helium, 252; track in Wilson chamber, 149–50
alpha ray, 55–6, 59. (*See also* alpha particle *and* helium)
aluminium, 168, 199
Amaldi, 203, 205
americium, 210
Anaxagoras, 19
Anaximander, 17, 18
Anaximenes, 17, 18
Anderson, 173–4, 176, 239, 240, 272, 292
annihilation of matter, 175, 177, 292, 316, 317
anti-baryon, 314
anti-gravity, 319
anti-matter, 315–20
anti-neutrino, 294, 297, 308–9, 314
anti-neutron, 292–3, 302, 312, 314, 315
anti-particle, 292–4, 303–4; and anti-matter, 315; and baryonic and electric charge, 315; and stability of matter, 312–15
anti-proton, 167, 293, 296, 302, 314, 315, 317, 319
argon, 121, 275
Aristotle, 17, 18, 19, 21, 23, 24
artificial radioactivity; discovery, 199; in heavy elements, 208–11; isomers, 201; neutron bombardment, 202–11; new elements, 210–11; uranium, 209–10, 270; uses of radio isotopes, 281–2
artificial transmutations, 195–8. (*See also* artificial radioactivity)
associated production of strange particles, 300–3
astatine, 210
Aston, 115, 158, 159, 160
atomic bomb, 272, 276–80
atomic number, 36, 64, 66; and isotopes, 115, 159–60, 169–70
atomic physics, 67, 68–113
atomic theories: after the Renaissance, 24–7; Avogadro's, 30–3; Dalton's, 27–30; Greek, 19–23
atomic weights, 32–4; chemical and physical scales of, 160; periodic law, 35–8
atoms; masses of, 38, 40, 157; dimensions of, 64–5
avalanche of ions, 139, 155
Avogadro, 31–3; Avogadro's Law, 31–2

Bacon, 24
barium, as fission product, 209–10
baryon, 290 *n*, 293, 295, 309; and resonances, 311
baryonic charge, 313–15
Becker, 164, 165
Becquerel, 51–2, 56,·128
berkelium, 210
Bernardini, 239
Bernoulli, 26
Berthollet, 27

324　　　　　　　　　Index

beryllium; artificial disintegration of, 164–5, 203; cosmic radiation, 232, 233, 244; moderator, 205, 273
beta decay, 117, 188–9, 191–2, 204, 209, 294, 295, 297, 298, 311; and parity, 305–8
beta ray, 55–6, 142
betatron, 215, 219–20
Bethe, 266
binding energy, 172–3, 178, 179; and nuclear energy, 248–53
black body radiation, 72, 74
Blackett, 148, 174, 239
Blau, 244
Bocciarelli, 239
Bohr, 64, 103, 107, 109; correspondence principle, 78–9, 80–1, 88; drop model, 185, 207, 208, 210; theory of hydrogen atom, 74–8, 80–1, 82, 83, 87–8, 96; wave-particle dualism, 100, 110–12
Bohr–Sommerfeld theory, 80–1, 87–8, 89, 206
Boltwood, 53
Boltzmann, 41, 43, 108, 290 n
Born, 99, 100–1, 103, 109
Bose–Einstein statistics, 290
boson, 287, 290, 295
Bothe, 164, 165, 201
Bragg, 95
braking radiation, 259, 260
de Broglie, 74, 104, 107, 112; and wave mechanics, 89, 95–101, 104
de Broglie wave, 96, 98–101
bromine, radioactive, 201
Brown, 38
Brownian motion, 38–9
Bruno, Giordano, 24
bubble chamber, 153–4
Bunsen, 71
Burbridge, 319
Butler, 299

Calder Hall, 280–1
californium, 210
canal ray, 49, 69
Cannizaro, 33
carbon; and Conversi–Pancini–Piccioni effect, 241; and cosmic rays, 232; and gamma ray bombardment, 252; as moderator, 205; from radioactive nitrogen, 200; in stars, 265–6
carbon cycle, 265–6
cascade process, 244
cathode ray, 46–7, 69
causality; and radioactivity, 122–3; and uncertainty, 107–11
Cerenkov counter, 141, 143–5
Cerenkov effect, 143–5
Chadwick; and beta decay, 189; and neutron, 165, 166, 190; and nuclear photo-electric effect, 198; and proton emission, 163
chain reaction, 268–9, 271–2, 277
Chamberlain, 293
charge independence, principle of, 181
Chauson, 48
chemical separation of artificial radioactive elements, 202
Cicero, 22
circular accelerators, 218–25
classical physics, 108–9
Classical Theory of Quanta, 89. (See also Bohr–Sommerfeld theory)
Classical Theory of Radiation, 74; and Rutherford's model, 70, 71
clouds, 145–6
cloud chamber, see Wilson chamber
Cockroft, 197, 215
cohesion forces, 41
Complementarity Principle, 110–12
Compton, 133
Compton effect, 132, 133–4, 150, 177, 178, 244
conservation laws; baryonic charge, 313–15; electric charge, 295, 313, 314; energy, 188–9, 191, 291, 295, 301, 304, 307; matter, 26–7, 301; momentum, 295; parity, 303–8; strangeness, 301–3, 307–8
Conversi, 241, 242, 277
Conversi–Pancini–Piccioni effect, 241
copper, and neutron bombardment, 203
corpuscular theory of light, 90–1
Correspondence Principle, 78–9, 88
cosmic radiation, 128–9; antiproton in, 293; constitution of primary radiation, 231–3; discovery, 226; east–west effect, 231; electrons in, 48; energies of, 226–7; intensity,

Index

cosmic radiation—*continued* 228–31; investigation of, 227; latitude effect, 229–31; mesons in, 238–44; origins of, 233–8; positrons in, 173–4, 238–9; secondary, 226, 229, 238–9, 244–7
Cowan, 191, 308
counters, 135–45, 155; in coincidence, 139–41
critical size, 271–2, 277–8
Crompton, 104
Crookes, 45
Crookes' tube, 46, 49, 69
Curie, Irene, *see* Joliot–Curie
Curie, Marie, 52–4, 175; and radioactivity, 53, 125, 141
Curie, Pierre, 52–4; and radioactivity, 53, 125, 127, 141
curium, 210
cyclotron, 221–3; frequency-modulated, 224

Dalton, 27–30, 31, 33, 41
Danby, 309
Davisson, 97
Debierne, 53
Dematerialization, 175, 177, 292, 316, 317
Democritus, 21–3, 24, 112, 122, 167, 320
deuteron, 158, 168, 169; and accelerating machines, 213, 223, 224; bombardment with, 195, 197–8, 201; fusion, 254–63; mass defect, 248–50; proton-proton process, 265–6; rest energy, 214
deuterium, 158, 168, 198, 205; binding energy, 248–9; fusion, 254–63; plasma, 257–63
diffraction; of light, 91–2; of electrons, 97
Diogenes, 21
Dirac, and anti-particles, 303, 316; and electromagnetic interactions, 297, 298; and operator method, 89, 103; and positron, 175–6, 292; and relativistic theory of electron, 83, 163, 175–6, 287, 292
displacement law, 117
drop model of nucleus, 183, 185, 207–8, 210

Dumas, 32–3
Dyson, 283

Earth; magnetic field, 229–30; internal heat, 126–8
east–west effect, 231
Eddington, 265
Einstein; and Bose–Einstein statistics, 290 n; and Brownian motion, 39; and determinism, 109; and equivalence of mass and energy, 171–2, 175, 248–9, 317; and kinetic theory, 41; and photoelectric effect, 92–5; and Quantum Theory, 72, 73, 78, 81, 90, 92–5, 96; and relativity, 103, 171
einsteinium, 210
eka-boron, eka-aluminium and eka-silicon, 36
electrolysis, 45
electromagnetic interactions, 291–2, 295, 297, 298, 302, 304, 309
electromagnetic radiation; and complementarity principle, 110–11; corpuscular theory, 90–1, 92, 94–5; ionization, 68; propagation, 57–8; quantum theory, 72–4, 75–83, 86–7, 88, 92–5; spectra, 71, 78; uncertainty principle, 106; wavelengths, 57–9; wave theory, 90–2, 94–5, 283. (*See also* gamma ray, X-ray, *etc*.)
electron; and accelerating machines, 213, 219–20, 223–4; atomic weight, 160; baryonic charge, 314; Bohr's model, 76–9, 81, 82, 167, 183; cathode ray, 46; charge, 47–8, 64, 287, 313; cosmic rays, 48, 238, 244, 245, 246, 247; Dirac's theory, 83, 163, 175–8, 287; discovery, 45–6; electromagnetic interactions, 291–2; electron-positron pair, 175–8, 292; exclusion principle, 83–7; ionization, 68–9, 131, 132–3, 135, 149, 150, 158; mass, 47, 48, 287; materialization electron, 175; meson decay, 194, 240–1; photo-electric effect, 93–4, 177–8; plasma, 257, 258, 260; positive, *see* positron; radioactivity, 55–6, 117, 127, 186,

326 Index

electron—*continued*
188-9, 191-2, 213, 270; rest energy, 214; Rutherford's model, 63, 65-7, 69-70; Sommerfeld's theory, 80-1; spin, 82-3, 287-90; stability, 312-13; statistics, 289; Thomson's model, 50-1; track in Wilson chamber, 149; uncertainty principle, 104-5; wave theory of, 95-101
electron layers or shells, 80, 84-7
electron volt, 162
electrostatic forces in nucleus, 179, 251
element (chemical), 26
emanation; of thorium and actinium, 120; of radium, *see* radon
Empedocles, 19, 21
endoenergetic reactions 251-2
Epicurus, 24
Epstein, 229
exchange forces, 182, 185, 192
excitation of atoms, 131, 134, 135, 141
Exclusion Principle, 84-5, 176, 182, 289, 290 n, 295, 309
exoenergetic reactions, 251-2
extensive shower, 244, 245, 246

Faraday, 45, 283
fermi (unit of length), 179
Fermi, 179, 190, 285; and beta decay, 191-2, 294, 297, 298, 311, 312; and Fermi-Dirac statistics, 289, 290 n; and meson, 241; and nuclear reactor, 272, 273; and neutron absorption lines, 205-6; and neutron bombardment, 202-3, 208, 209, 268; and positron, 192; and theory of cosmic rays, 234-7
Fermi-Dirac statistics, 289, 290 n
fermions, 289, 290, 292, 295
fermium, 210
fission, 209-10, 253, 267-72; and atomic bomb, 276-80; and nuclear reactors, 272-6
francium, 211
Franck, 80
frequency-modulated cyclotron, 223, 224

Fresnel, 90-2
Frisch, 210
fusion, 253, 254-67, 269; in H-bomb, 278-9
fusion reactor, 263

Gaillard, 309
Galileo, 23, 24
gamma ray; artificial radioactivity, 164-5, 203, 206, 208, 270; atom bomb, 279-80; bombardment with, 195, 198, 201, 252; Compton effect, 132, 133-4, 150, 178, 244; cosmic rays, 129, 226, 238, 244, 245, 246; discovery, 55-6; electron-positron pairs, 175-7, 178, 292; meson, 194; natural radioactivity, 55-6, 117, 127, 130, 185-6, 269, 281; nature of, 59, 132; scintillation, 142; Wilson chamber, 151
Gamow, 187-8
gaseous diffusion process, 277
Gassendi, 24
Gay-Lussac, 30-1
Geiger, 63, 64
Geiger counter, 138-9, 148-9
Gell-Mann, 285, 292, 296, 301
Gentner, 201
Germer, 97
Glaser, 153
gold and neutron capture, 205-6
Goldhaber, 319
Goldharbour, 198
Goldstein, 49
Gorgias, 20
Goulianos, 309
graphite, 273
gravitational forces in nucleus, 178, 179
gravity, 307, 319
Grimaldi, 90-1

Hahn, 201, 209, 210
half-life, 121-2, 188
halogens, electron layers of, 85
heavy hydrogen, *see* deuterium
heavy water, 168, 273
Heisenberg; and isotopic spin, 301-2; and matrix mechanics, 89, 100,

Index

Heisenberg—*continued*
101, 103-4, 284; and nuclear forces, 181-2, 192; and uncertainty principle, 99, 104-7, 109
helium; atomic structure, 66, 163; binding energy, 173, 178, 249, 250; cooling reactors, 275; cosmic rays, 232; electron orbits, 84-5; fusion, 254-5; radioactivity, 59-60; stellar energy, 265-7; transformation of lithium, 252. (*See also* alpha particle)
Helmholz, 45, 46
Heraclitus, 18
Hertz, 80, 92, 283
Hess, 226
Heyn, 203
Hiroshima, 278
Hoffman, 53
Hoyle, 319
Huyghens, 90-1
hydrogen; and atomic weights, 33-4, 115, 116, 160; Bohr's theory, 75-8, 80, 88; exclusion principle, 83, 84; in interstellar space, 317; isotopes, 158, 168-9 (*see also* deuterium *and* tritium); mass of atom, 40; molecule, 31, 32; nucleus, 116, 168 (*see* proton); Rutherford's model, 66, 75; stellar energy, 265-7
hydrogen bomb, 278-9, 280
hyperon, 311

ideal experiments, 102
indicators, 282
infra-red radiation, 58
interference, 90, 91
interstellar matter, 235
ionic binding, 85-6
ionization, 68-9; of atmosphere, 128-9, 226; of interstellar matter, 235; by particles, 131, 134; by photons, 132-3, 134
ionization chamber, 135-6, 141; rapid, 137; liquid, 137, 138; crystal, 137-8
ions, 68, 69; and mass spectrograph, 158-9
iron; and cosmic rays, 232; and meson absorption, 241; from radioactive manganese, 209

isomers, 201-2
isotopes, 114-16, 157-9; and mass number, 160, 170; uses of radioisotopes, 281-2
isotopic spin, 301-2

Janossy, 245
Jeans, 122
Joliot-Curie (Irene Curie and F. Joliot), 164-5, 199-200, 201
Joly, 128
Jordan, 89, 103

K-particle, 299, 300, 302, 311
Kelvin, Lord, 127, 128
Kerst, 219
kindling temperature of plasma, 260
kinetic theory, 26, 40, 44
Kirchhoff, 71
Kunze, 239

Laborde, 127
lambda particle, 299, 300, 302
Laplace, 26
latitude effect, 229-31
Lattes, 242
Lavoisier, 26-7, 38
law of constant proportions, 27, 29
law of multiple proportions, 27-30
law of octaves, 35
Lawrence, 197, 221
lead, and radioactive decay, 120, 124-5
Lederman, 309
Lee, 304, 305, 311, 320
Leprince Ringuet, 193, 243, 247
Liebnitz, 303
Lemaître, 229
leptons, 290 *n*, 295, 309
Leucippus, 20-3, 24
light, 90-5; visible, 54, 86, 87; quantum, *see* Photon. (*See also* electromagnetic radiation)
linear accelerators, 215-18
lithium, 66, 85, 249; in cosmic rays, 232, 233, 244; transformation into helium, 197, 252

Livingstone, 221
Lucretius, 24
lutetium, 121

magnesium and alpha particle bombardment, 200
magnetic bottle, 261-3
magnetic field, effect on charged particle, 158, 218-19
Majorana, 182, 192
manganese and neutron bombardment, 209
Marsden, 63, 64
Marshak, 166, 242
mass defect, 170-3, 248-9
mass number, 159-60, 170
mass spectrograph, 115, 158-9, 277
materialization, 132, 134, 175
materialization electrons, 175
Matrix Mechanics, 89, 101-4, 284
Maxwell; and electromagnetic theory, 92, 193, 283, 291; and kinetic theory, 41, 43, 108
Mayer, Marie, 184
McMillan, 223
mean binding energy per nucleon, 249-51
mean free path, 44
Meitner, Lisa, 209, 210
Melixes, 19
membrane Wilson chamber, 148
mendelevium, 210
Mendeliev, 35-8, 309, 310
mesic atom, 241
meson, 167, 295, 309; Conversi-Pancini-Piccioni effect, 214; cosmic radiation, 238-42, 245, 246, 293; decay, 240, 243-4; Yukawa's hypothesis, 192-4, 239, 241, 293. (See also mu meson and pi meson)
meson field, 194
meteorites, 126, 128
microscope, 212
Millikan, 47
mirror invariance, 304
Misty, 309
mixed shower, 244, 245, 246
moderators, 205, 272-3
molecule, 30-3
molybdenum, 267
Moon, 205

Morrison, 303, 308
Moseley, 64
mu meson, 296-7, 298; cosmic rays, 242-3, 245, 246, 247; decay, 243, 297
muon, see mu meson

Nagasaki, 278
natural radioactivity; age of rocks, 125-6; alpha decay, 186-8; in atmosphere, 128-9; beta decay, 188-92 (see also beta decay); detection of radioactive particles, 130-56; discovery of, 48, 51-4; displacement law, 117; gamma ray, 185; half-life, 121-3; heat of Earth, 126-8; investigation of, 54-6, 59-60; isotopes, 115-16; radioactive substances in nature, 125; nucleus, 67, 114, 161; radioactive equilibrium, 123-5; radioactive families 119-21, 124-5; theory of (Rutherford and Soddy), 117-18
Neddermayer, 239
neptunium, 210, 270
neutrino, 167, 244, 293, 295; and beta decay, 189-92, 294, 305, 308-9, 314; and meson decay, 194, 240, 243, 246, 296-7, 309
neutron, 150, 213, 308; and antineutron, 292-3, 312, 315; baryonic charge, 314, 315; bombardment with, 185, 195, 196, 202-3, 208-11; cosmic rays, 237, 238, 245, 247; decay, 191-2, 237, 294; and deuteron, 248-9; discovery, 163-6, 190; emission under alpha bombardment, 164-5, 200; fission, 209-10, 267-9; fusion, 254, 255, 279; and nuclear forces, 179, 180-2, 194, 251, 293; and nuclear photoelectric effect, 198, 201, 252; and nucleus, 167-73, 184-5, 186, 188, 268, 285, 287; resonances, 311; slow neutrons, 203-8; spin, 288; statistics, 289; strangeness, 301, 302
neutron absorption lines, 205-8, 271
neutron-proton diagram, 169-71, 184, 267-8
Newlands, 35

Index 329

Newton; and atomic theory, 24; and determinism, 122; and theory of light, 90–1, 95; and theory of planets, 26, 108
Nishijima, 301
nitrogen; alpha particle bombardment of, 162–3, 195, 197; in carbon cycle, 266; in cosmic rays, 232, 233; transmutation of boron, 200
noble gases, 38; electron layers of, 85
Northern Lights, 229
nuclear bomb, see atomic bomb
nuclear energy; and binding energy, 248–51; fission, 253, 267–78; fusion, 253–67, 278–9; nuclear reactions, 251–2; peaceful uses, 280–2
nuclear field, 193–4, 293, 297. (See also meson)
nuclear forces; charge independence, 181; meson, 192–4, 239, 240, 242, 243, 293–4, 297, 298; radius of action, 179; 'repulsive core', 182; saturation, 181–2
nuclear interactions, 293–4, 295–9. (See also Meson)
nuclear models, 182–5. (See also drop model and shell model)
nuclear photo-electric effect, 198, 201, 252
nuclear physics, 67, 157
nuclear propulsion, 281
nuclear reactions; endoenergetic, 251–2; exoenergetic, 251–2; symbolism for, 196–7
nuclear reactor; control of, 274; cooling of, 275; moderators, 272–4; types of, 276; uses of, 280–1
nuclear shell, 183–5
nuclear transformations, 161–3. (See also artificial and natural radioactivity)
nucleon, 167, 192, 193; mean binding energy per nucleon, 249–51. (See nuclear forces)
nucleus; bombardment of, 161, 162–166, 195–211; charge, 64; fission, 209–10, 253, 267–72; fusion, 253, 254–67; isotopes, 116, 160; radioactivity, 117–18, 130, 185–92; Rutherford's model, 62–7; size, 64–5; structure, 167–9, 182–5. (See also nuclear forces, etc.)

Occhialini, 148, 174, 239, 242
operational definitions, 102–3
Operator Method, 89
optical model of nucleus, 183, 185
oxygen; and atomic weights, 34; and carbon cycle, 266; and cosmic rays, 232; and reactors, 275; and transmutation of nitrogen, 195, 197

Pais, 300
Pancini, 226, 241, 242, 297
paraffin and slow neutrons, 204–5
parallel plate counter, 155
parity, 303–8, 311, 316–17, 320
Parmenides, 18, 19, 20, 21
Pauli, 84, 307; and exclusion principle, 84–5, 176, 182, 289, 290 n, 295, 309; and neutrino, 189–90, 191, 294
penetrating radiation, 226. (See also cosmic radiation)
penetrating shower, 244, 245, 246
Periodic System, 35–8, 64, 84–5, 87, 115, 169, 183, 309
Perrin, 39
Persico, 88, 89
photon; as anti-particle, 293; baryonic charge, 313–14; electron-positron pairs, 134, 175, 176–8; emission by nucleus, 185–6; meson decay, 296; Quantum Theory, 73–4, 92–4, 95, 96, 193. (See also Compton effect, electromagnetic interactions, gamma ray, photo-electric effect, etc.)
pi meson, 293–4, 297, 298, 299, 300 n, 302, 311; cosmic rays, 242–3, 245, 246, 247; decay, 243, 244, 296, 309
Piccioni, 241, 242, 297
pilot wave, 99
pinch effect, 261–3
pitch-blende, 53, 125, 269
pion, see pi meson
phosphorus; artificially radioactive, 201; and tracer elements, 282

photo-electric effect, 92–4, 132, 134, 150, 177, 235, 244; and quantum theory, 73
photographic emulsion method, 151–2
photomultiplier, 142–3
Planck, and Quantum Theory, 72–4, 75, 78, 86, 90, 92, 96–7, 290 n
Planck's constant, 73, 82, 89, 94, 95, 96–7, 106, 107, 111–12, 288
plasma, 256–7; of deuterium gas, 257–63; energy losses, 259–61; kindling temperature, 260; magnetic bottle, 261–3; pinch effect, 261–3
Plato, 17, 18, 20, 22
Plücker, 46
plutonium, 210, 270, 271, 276, 277
Poincaré, 71
polonium, 53, 131
Pontecorvo, 203, 309
Poole, 128
positron, 167, 293; and anti-matter, 315, 317; baryonic charge, 314; carbon cycle, 266; cosmic rays, 173–4, 238, 239, 244, 245, 247; discovery, 173–4; electron-positron pair, 175–8, 292, 313; mu meson, 243, 297; nuclear disintegration, 201, 203; proton, 192, 308, 312
Post, 257
Powell, 242
probability, 81; and probability waves, 99, 100–1
promethium, 211
proportional counter, 139
Protagoras, 20
proactinium, 267
proton, 66, 116, 159, 163, 164–5; atomic weight, 160; and anti-proton, 292–3, 296, 311, 317, 319; baryonic charge, 313–14, 315; bombardment with, 195, 197, 201, 252, 256; carbon cycle, 266; cosmic rays, 229, 231–3, 235–8, 245–7; detection, 142, 151; emission by nucleus, 162–3, 195, 197, 198, 199, 203, 207–8; and fission, 210; and fusion, 254, 256; and neutron, 191–2, 204, 249, 301–2, 308–9; and nuclear forces, 178–82, 192–4; and pion, 294, 296, 300 n; rest mass, 214; spin, 288; statistics, 289; strangeness, 302; structure of nucleus, 166–70, 172–3, 183–5, 251, 268
proton-synchrotron, 224–5
Proust, 27
Prout, 33–4, 116, 161

Quanta, Classical Theory of, 89. (See Bohr–Sommerfeld theory)
quantum; of energy, 72, 96 (see also Quantum Theory); of light, see photon and gamma ray
Quantum Electrodynamics, 291–2. (See also electromagnetic interactions)
Quantum Mechanics, 89, 123, 183, 284; wave equation of, 104. (See also Wave Mechanics, Matrix Mechanics and Operator Method)
Quantum Theory, 72–4, 88–9; and complementarity principle, 110–11; and Compton effect, 133; and correspondence principle, 79; and electromagnetic interactions, 291; and electron spin, 82–3, 287–8; and hydrogen atom, 75–8, 79–81, 86–8; and photo-electric effect, 92–5; and probability, 81; and uncertainty principle, 104–7; and wave-particle dualism, 96–7, 193

radio isotopes, uses of, 281–2
radio waves, 58
radioactivity, see artificial and natural radioactivity
radium; and alpha particle, 59–60, 118; discovery, 48, 53, 125; disintegration, 118–19; emanation of, see radon; gamma ray, 59; half-life, 121; heat emitted, 127; isotopes, 53, 118–19, 123–4
radon, 118, 119, 123, 124, 203
Rasetti, 203
refraction, 90–1
Reines, 191, 308
relativistic mechanics, 80, 81, 214
relativistic theory of electron, 83, 163, 175–6, 287, 292

Index

relativity, 99, 103, 303, 319
relay counter, *see* Geiger counter
repulsive core of nuclear forces, 182
resonances, 311
rest mass (rest energy), 172, 188, 214, 287
Rochester, 299
rocks, ages of, 125–6
Ronchi, 90
Rosenbaum, 285, 292, 296
Rossi, 244, 245
Royds, 59–60
rubidium, 121
Rutherford, and alpha particle, 59–60; and collision experiments, 60–3, 64, 162–3, 164, 195; and model of atom, 51, 62–7, 69–71, 74–5, 283; and neutron, 165 n; and scintillation method, 141; and study of radioactivity, 54, 55, 117
Rydberg constant, 78

Salam, 307
samarium isotope, 121
saturation of nuclear forces, 181
scandium and fission, 267
Schiff, 319
Schrödinger, 23, 97, 104
Schwarz, 309
scintillation counter, 141–3
secondary ionization, 149
second law of thermodynamics, 102
Segrè, 203, 293
shell model of nucleus, 183–5, 206–7
shower, 174, 244, 246, 299
sigma particle, 302
silicon and artificial radioactivity, 199, 200
silver and neutron bombardment, 204
simultaneity, 103
'sixty years law', 283-4, 312
slow neutron, 203–8, 270–1
Smoluchowski, 39
Socrates, 20, 21
Soddy, 53, 115, 121, 179; theory of radioactive transformations, 117
Sommerfeld, 80; Sommerfeld's conditions, 80–1. (*See also* Bohr–Sommerfeld theory)
Spark chamber, 155–6
spectroscopy, 45, 71
spectrum, 71; and Bohr's theory, 78–9, 80–1; and electron spin, 82; and Schrödinger's calculations, 97; X-ray, 64
spin, 287–9; of electron, 82–3, 287, 288
spinthariscope, 142
stability of matter, 312–15; of nucleus, 170, 184, 250
star (cosmic radiation), 244, 245, 246
stars, energy of, 236–7
Stas, 34
statistics, 288–90
Steinberger, 309
Stoney, 45, 46
Störmer, 229
strangeness, 301–3
strange particles, 299–301, 302, 304
Strassmann, 209, 210
Strauss, 53
strong interactions, 297–8, 299, 300, 301, 302, 309
sun; and cosmic rays, 232, 233; energy of, 263, 266–7
supernovae, 234
synchrotron, 223–4
Szilard, 273

Tatel, 228
technetium, 211, 267
Teller, 241
Thales, 17, 18
thermonuclear reaction, *see* fusion
thermonuclear reactor, 169, 263
Theophrastes, 18
Thomson, and atomic model, 50–1, 62–3; and cathode rays, 46; and electron, 158, 159
thorium; fission, 267; half-life, 121; isotopes, 121, 198
thorium family, 119–20
Tillman, 205
tracer elements, 282
track detectors, 145–56
trans-uranic elements, 210–11
tritium, 169, 198, 254, 263; fusion with deuterium, 255, 261; in hydrogen bomb, 279
triton, 169. (*See also* tritium)

Uhlenbeck, 82–3, 287
Uncertainty Principle, 99, 104–7, 109, 111–12
'universon', 319–20
uranium; alpha decay, 187; atomic bomb, 277; beta decay, 189, 190; binding energy, 178; discovery of, 269; discovery of radioactivity, 51–2; fission, 209–10, 267–9, 270–2, 277; isotopes, 119, 269–70, 277; mass of atom, 40; nucleus, 66, 170; reactors, 272–4; in rocks, 124–6; trans-uranic elements, 210–11

Vallarta, 229
Van Allen, 228
Van der Broeck, 64
Vekseler, 223
Villard, 57
virtual process, 291, 293, 294

Walton, 197, 215
Wambacher, 245
Watoghin, 245
wave theory of light, 74, 90–2, 93, 94–5
Wave Mechanics, 89, 95–101, 104, 111, 193; and alpha decay, 187–8
weak interactions, 297–8, 300, 312; and neutrino physics, 308–11; and parity, 304–8, 311, 316–17, 320; and strangeness, 302, 303. (*See also* beta decay)
Weigand, 293
Weisskopf, 180, 241, 307
Weiszächer, 202
Wideroe, 217, 219
Wiener, 38–9
Wigner, 181
Wilson, 145, 146
Wilson chamber, 145, 146, 151, 152, 153–4; membrane, 148
wire chamber, 156
Wu, 307

X-ray; braking radiation, 260; Compton effect, 133; discovery, 48, 51; frequency, 59; penetrating power, 54–5; photo-electric effect, 178; spectra, 64
xi particle, 302

Yang, 301, 304, 311, 320
Young, 90, 91
Ypsilantis, 293
Yukawa; and meson, 193–4, 239, 240–2, 243, 293, 297; Yukawa's process, 295–9. (*See also* meson)

Zinn, 272
zirconium from fission, 267
Zworykin, 48